HISTORY
OF THE
UNITED STATES
BOTANIC GARDEN

1816–1991

by
Karen D. Solit

PREPARED BY THE ARCHITECT OF THE CAPITOL
UNDER THE DIRECTION OF
THE JOINT COMMITTEE ON THE LIBRARY

CONGRESS OF THE UNITED STATES

WASHINGTON   1993

For sale by the U.S. Government Printing Office
Superintendent of Documents, Congressional Sales Office, Washington, DC 20402
ISBN 0-16-040904-7

The Summer Terrace Display, an annual event sponsored by the Botanic Garden since 1973.

FOREWORD

The Joint Committee on the Library is pleased to publish the written history of our Nation's Botanic Garden based on a manuscript by Karen D. Solit. The idea of a National Botanic Garden began as a vision of our Nation's founding fathers. After considerable debate in Congress, President James Monroe signed a bill, on May 8, 1820, providing for the use of five acres of land for a Botanic Garden on the Mall. The *History of the United States Botanic Garden,* complete with illustrations, traces the origins of the U.S. Botanic Garden from its conception to the present.

The Joint Committee on the Library wants to express its sincere appreciation to Ms. Solit for her extensive research and to Mr. Stephen W. Stathis, Analyst in American National Government with the Library of Congress' Congressional Research Service, for the additional research he provided to this project.

Today, the United States Botanic Garden has one of the largest annual attendances of any Botanic Garden in the country. The special flower shows, presenting seasonal plants in beautifully designed displays, are enjoyed by all who visit our Nation's Botanic Garden.

The United States Botanic Garden, located at the foot of the Capitol, will be celebrating its 175th Anniversary in 1995. The *History of the United States Botanic Garden* is a fitting tribute to this important occasion.

Claiborne Pell
Vice Chairman,
Joint Committee
on the Library.

Charlie Rose
Chairman,
Joint Committee
on the Library.

United States Botanic Garden
245 First Street, SW
Washington, DC 20024

Botanic gardens in the modern sense have played an important role in the history of civilization since the first one was founded in the sixteenth century. The United States Botanic Garden is no exception. With a history that parallels the development of this country, it is the oldest continually operating botanic garden in the United States.

This volume is a fascinating chronicle of the development of the institution now known as the United States Botanic Garden. The idea of a national botanic garden started with the Columbian Institute and was revived by the Wilkes Exploring Expedition. In an age where adventures have become commonplace, it is necessary to appreciate the courage of the early explorers like Charles Wilkes. Their true risk-taking brought botanic curiosities to the attention of the nation through the U.S. Botanic Garden, which maintains some of the original collections.

Appreciation goes to Karen Solit, the author of the original manuscript, and to the staff of the Congressional Research Service, the Curator's Office for the Architect of the Capitol, and the U.S. Botanic Garden, who assisted in the preparation of the text. Unless otherwise credited, the photographs are from the records of the Office of the Architect of the Capitol.

As we approach the 175th anniversary of the founding of the United States Botanic Garden, this document gives us a perspective on the past and renewed dedication to the future of this special public resource.

George M. White, FAIA
Architect of the Capitol
Acting Director, U.S. Botanic Garden

CONTENTS

FRONTISPIECE ii
FOREWORD iii
LETTER OF ACKNOWLEDGEMENT v
PREFACE xi

I. THE COLUMBIAN INSTITUTE FOR THE PROMOTION OF ARTS AND SCIENCES (1816 - 1837) AND THE FIRST BOTANIC GARDEN 1
   Founding 1
      Objectives of the Institute 2
   Congressional Charter 5
   First National Botanic Garden 5
      Design of the Institute's Garden 6
      Plant Collection and Distribution 7
      Plants Cultivated in the Institute's Garden 11
      Maintaining the Garden 11
   Other Activities of the Institute 12
      Beginnings of a National Herbarium 12
      Presentation of Scholarly Papers 12
      Collection of Objects 13
   Reasons for the Institute's Failure 13

II. THE WILKES EXPEDITION (1838 - 1842): A SECOND CHANCE FOR THE NATIONAL BOTANIC GARDEN 17
   Expedition's Scientific Corps 19
   Anything But Smooth Sailing 19
   Expedition's Accomplishments 20
   Expedition's Record: A Permanent Legacy 21
   America's Response to the Expedition 22
   Disposition of the Plant Material Collected 23
   Need for a New Location 28
   Wilkes Plaque 29

III. THE UNITED STATES BOTANIC GARDEN: NEW FACILITIES IN A FAMILIAR SETTING 31
   The William Smith Years (1853 - 1912) 32
   The Garden in the Latter Half of the Nineteenth Century 36
      The Bartholdi Fountain 36
      Conservatories 38
      Botanical Collections 38

A New Plan for the Capital City and Its Botanic
   Garden   39
End of an Era   42
The George Wesley Hess Years (1913–1934)   42
Debate Over Relocation   46

IV. THE PRESENT UNITED STATES BOTANIC
   GARDEN   49
   A New Location and a New Conservatory   49
   Development of the Botanic Garden Park   53
   New Growing Areas   53
   Administration of the Garden   55
   Current Functions of the Garden   58
   Future Directions of the Garden   62
   Conclusion   63

V. A SELECTED BIBLIOGRAPHY   65

VI. APPENDIXES

   Appendix 1 Circular Letter and Report of 1827   71
   Appendix 2 Alphabetized Version of William Elliot's List of Plants in the Botanic Garden of the Columbian Institute, Prepared in 1824   79
   Appendix 3 Experimental Garden of the National Institute, 1844   87
   Appendix 4 Memorial Trees Planted on the Grounds of the First U.S. Botanic Garden   89
   Appendix 5 Species Included in "A Catalog of Plants in the National Conservatories," Prepared by William R. Smith in 1854   95
   Appendix 6 Keim's Description of the Botanical Collection, 1875   99
   Appendix 7 Plants Collected During the Perry Expedition of 1852 to 1855 and Donated to the Botanic Garden   103
   Appendix 8 Plants Growing Inside the Conservatory and on the Grounds of the First Botanic Garden During the George W. Hess Years (1913–1934)   105

Appendix 9 Plants for Congressional Distribution Through the Botanic Garden, 1930   109

ILLUSTRATIONS

Summer Terrace Display   Frontispiece   ii
Lieutenant Charles Wilkes   16
Vessel Fern   27
William R. Smith   34
Bartholdi Fountain   37
Laying of the Cornerstone   48
Dome of the Original Main Conservatory   50
Inside the Subtropical House   52
Orchid Display in the U.S. Botanic Garden Conservatory   60

PREFACE

Botanic gardens are living museums. They preserve the world's flora in a protected environment while at the same time offering scientists, students, and the general public the opportunity to study a wide variety of plant life in an aesthetic and educational setting. The United States Botanic Garden is one of the nation's oldest such institutions and one of the few of its kind that is primarily funded by the Federal Government.

The creation of the United States Botanic Garden is an integral part of our horticultural heritage. Its history is almost as old as the Nation and is an invaluable source of information for those interested in the role and functions of American public gardens; in the collection, cultivation, and display of the world's flora; and in the history of the nation's capital.

A botanic garden at the seat of government was the dream of several of America's earliest statesmen, including George Washington, Thomas Jefferson, and James Madison. That dream first became a reality when the first botanic garden was established in 1820 under the auspices of the Columbian Institute for the Promotion of Arts and Sciences.[1]

The Columbian Institute, through its garden, sought to collect, cultivate, and distribute plants from all over the world. During the garden's nearly two-decade existence, it succeeded in amassing a diversified plant collection and distributing economically useful plants throughout the nation. A continual lack of funds, as well as several other problems, however, prevented the Institute from ever fully developing its garden. The organization stopped holding meetings in 1837, and its botanic garden ceased functioning soon thereafter.

The idea of a national botanic garden was revived in 1842 when the United States Exploring Expedition to the South Seas, led by Lieutenant Charles Wilkes, brought to Washington a collection of living plants from around the

---

[1] Richard Rathbun, The Columbian Institute for the Promotion of Arts and Sciences, Washington: U.S. Govt. Print. Off., 1917, pp. 37–39.

globe. Initially, these plants were placed in a greenhouse specifically erected for the collection behind the Old Patent Office Building in Washington. There they flourished until 1849, when an addition to the Patent Office made it necessary to find a new location for the plants.

Congress first approved construction of a new greenhouse the following year at the west end of the Capitol grounds on the exact site occupied by the Columbian Institute's garden. The Botanic Garden, officially named in 1856, prospered in its new location, receiving widespread public attention for its diversified plant collection and unique seasonal displays.

Early in the 1930s, the Botanic Garden was relocated, as part of an extensive Washington beautification plan, to its present site at Maryland Avenue and First Street, Southwest, directly south of the original site. Today, a century-and-a-half after the founding of its earliest incarnation, the United States Botanic Garden continues to carry on the horticultural and educational objectives first espoused by the Columbian Institute for the Promotion of Arts and Sciences.

CHAPTER I

THE COLUMBIAN INSTITUTE FOR THE PROMOTION OF ARTS AND SCIENCES (1816–1837) AND THE FIRST BOTANIC GARDEN

The Columbian Institute for the Promotion of Arts and Sciences, in creating the first national botanic garden in America, established the precedent for the present United States Botanic Garden at the foot of Capitol Hill. What link today's Botanic Garden with the one started by the Columbian Institute in 1820 are a similarity of purpose, a common site, and the national character of the two organizations.

Although the development of a botanic garden was the institute's most fully realized objective, this was far from its only purpose or achievement. An examination of the Institute's accomplishments is useful for understanding the setting in which the first national botanic garden was conceived.

FOUNDING

The Columbian Institute evolved from the Metropolitan Society founded on June 15, 1816, which, it was hoped, would become the "Washington equivalent of the American Philosophical Society in Philadelphia."[2] Credit for the Society's formation goes to two of its founding members, Dr. Edward Cutbush of Pennsylvania and Thomas Law of England.[3] Dr. Cutbush, a Navy surgeon stationed in Washington, was the Society's most zealous exponent and the leading spirit behind its establishment. He authorized its organizational plans, was a member of the committee that drafted the successor Columbian Institute's constitution,

---

[2] Constance McLaughlin Green, Washington: Village and Capital, 1800–1878, 2 Vols. Princeton, N.J.: Princeton University Press, 1962, v. 1, p. 69.
[3] Biographical sketches of Edward Cutbush and Thomas Law are found in G. Brown Goode, The Genesis of the National Museum, in Annual Report of the Board of Regents of the Smithsonian Institute for the Year Ending June 30, 1891, Washington: U.S. Govt. Print. Off., 1892, pp. 276n–282n.

and then served as the Institute's first president.[4] It was Thomas Law, however, a leader in the intellectual life of the new capital city, who actually suggested the formation of such an organization.

The organizational plan approved by Cutbush, Law, and the other eighty-seven founding members of the Metropolitan Society called for the group to collect, cultivate, and distribute the various vegetable products of this country, as well as other nations, and to accumulate a collection of minerals found throughout the world. From the outset, the Society contemplated asking Congress if it might use two hundred acres on the mall to cultivate the plants it hoped to collect.

To accomplish this goal, the Society on June 28, 1816, selected a committee to draft a constitution. By early August, a constitution greatly expanding the organization's objectives had been agreed upon, and the name of the Society at that time was changed to the Columbian Institute for the Promotion of Arts and Sciences.[5]

Objectives of the Institute

The Institute's constitution contained several broad objectives:

Art. 1. The association shall be denominated the "Columbian Institute for the Promotion of Arts and Sciences"; and shall be composed of resident and honorary members.

Art. 2. The objects of the Institute shall be to collect, cultivate and distribute the various vegetable productions of this and other countries, whether medicinal, esculent, or for the promotion of arts and manufactures.

Art. 3. To collect and examine the various mineral productions and natural curiosities of the

---

[4] Rathbun, The Columbian Institute, p. 20. Subsequently Josiah Meigs, John Quincy Adams, John C. Calhoun, and Mahlon Dickerson served as presidents of the Institute. Ibid., p. 32.

[5] Ibid., pp. 10–11, 20; and Harold T. Pinkett, Early Agricultural Societies in the District of Columbia. Records of the Columbia Historical Society, 51–52, p. 39.

United States, and give publicity to every discovery which they may have been enabled to make.

Art. 4. To obtain information respecting the mineral waters of the United States, their locality, analysis and utility; together with such topographical remarks as may aid valetudinarians.

Art. 5. To invite communications on agricultural subjects, on the management of stock, their diseases and the remedies.

Art. 6. To form a topographical and statistical history of the different districts of the United States, noticing particularly the number and extent of streams, how far navigable; agricultural products, the imports and exports; the value of lands; the climate, the state of the thermometer and barometer; the diseases which prevail during the different seasons; the state of the arts and manufactures; and any other information which may be deemed of general utility.

Art. 7. To publish annually, or whenever the Institute shall have become possessed of a sufficient stock of important information, such communications as may be of public utility; and to give the earliest information, in the public papers, of all discoveries that may have been made by or communicated to the Institute.[6]

The activities deriving from these objectives, Dr. Cutbush told a large audience at Congress Hall on the evening of January 11, 1817, would bring numerous advantages to Washington. "It is true," Cutbush emphasized, "that in the infantile state of our city, we cannot boast of the possession of many, whose avocations have permitted them to devote their time to the cultivation of the sciences." Still, he felt confident that there were many people in Washington who possessed the "industry and an ardent desire to promote the objectives of the Institute." Those "minds, when allured to the contemplation of those objectives, aided by a botanical

---

[6] Rathbun, The Columbian Institute, p. 67.

garden, a mineralogical cabinet, a museum for the reception of natural curiosities, and a well selected library, will in a short period, be enabled to render essential services in many of the branches of knowledge embraced" by the Institute's constitution.[7]

Dr. Cutbush continued by telling those in attendance:

> . . . How many plants are there, natives of our soil, possessed of peculiar virtues, which would supersede the necessity of importing those that are medicinal or necessary for the operation of the dyer! . . .
>
> We have been peculiarly fortunate, my friends, that our association has commenced at the seat of government; where, through the representatives of the people, coming from the various sections of our country, of different climates and soils, whose minds are illuminated by the rays of science; and through scientific citizens and foreigners who visit this metropolis, we may reasonably expect, not only valuable communications, but various seeds and plants; hence, the necessity for a botanical garden where they may be cultivated, and, as they multiply, distributed to other parts of the Union. . . . The numerous grasses, grains, medicinal plants, trees, &c., which are not indigenous to our country should be carefully collected, cultivated and distributed to agriculturists.[8]

The advancement of knowledge for its own sake and the creation of a botanic garden with great aesthetic merit, however, were not the primary objectives of the Columbian Institute. Instead, the Institute perceived its role as being utilitarian—to collect, cultivate, and distribute plants through its garden that were potentially beneficial to the American people. The botanic garden was intended as a site for those plants.

---

[7] Ibid., pp. 12–13.

[8] Ibid., p. 13. G. Brown Goode credits Cutbush among others with opening the "way for the organization of the National Institute which . . . in turn . . . had an important influence toward shaping the course of the Smithsonian Institute." Indeed, in Goode's mind "the germ of the Smithsonian may be found in Cutbush's address." Goode, The Genesis of the National Museum, p. 279.

CONGRESSIONAL CHARTER

Two years later, on April 20, 1818, the Institute received a twenty-year congressional charter. The purposes of the Institute were reiterated in section four of that document:

> That the said corporation may procure, by purchase or otherwise, a suitable building for the sittings of the said institution, and for the preservation and safe-keeping of a library and museum; and also, a tract or parcel of land, for a botanic garden, not exceeding five acres: *Provided,* that the amount of real and personal property to be held by the said corporation shall not exceed one thousand dollars.[9]

FIRST NATIONAL BOTANIC GARDEN

After considerable lobbying, Congress on May 8, 1820, approved and President James Monroe signed into law a bill granting the Institute the use of five acres of land for a botanic garden.[10] Subsequently, three members of the Institute met with President Monroe to reach a decision regarding the selection of a parcel of land on the eastern end of the Mall, between Pennsylvania and Maryland Avenues, and extending from the base of the Capitol to the Tiber Canal at Second Street, Southwest.[11] Four years later, the allotment of land was extended to include an additional city block.[12]

Unfortunately, in the 1820s the Mall was desolate and unimproved. The northern and eastern sides of the Mall were low and swampy because the Tiber Creek, which flowed through the property, frequently flooded the area. This less-than-ideal site was one of the many difficulties the Columbian Institute had to overcome in establishing its botanic garden. In fact, the Mall was to remain unoccupied

---

[9] Rathbun, The Columbian Institute, p. 11.

[10] Ibid., pp. 12, 25–26; and 6 Stat. 247–248. The Institute's original handwritten minutes are found in Record of Proceedings of the Columbian Institute for the Promotion of Arts and Sciences (MC–358) Manuscript Division, Library of Congress (hereafter cited as Minutes of the Columbian Institute).

[11] Minutes of the Columbian Institute, May 29, 1820, pp. 65–66.

[12] 6 Stat. 316.

for nine years after the abandonment of the Institute's botanic garden in 1837 until the site for the Smithsonian Institution was selected between Ninth and Tenth Streets, Southwest.

Design of the Institute's Garden

Still, during its early years, the Institute was able to make several major improvements to the five-acre tract granted by Congress. First, a board fence five feet high was erected to enclose the garden, and then honey locust seeds (*Gleditsia tricanthos*) were purchased for planting along the inside perimeter of the fence. The intention was to remove the board fence after the honey locust had grown high enough to create a sufficient barrier. Later, the Institute unsuccessfully sought money from Congress to erect an iron or brick fence around the garden.

Several other improvements, however, were initiated and completed. The land was drained by the municipal commission and an elliptical pond, 144 by 100 feet, was formed with an island, 114 by 85 feet, in the center. A conduit was built from the pond to Tiber Creek, allowing the water to pass from one to the other. Four walkways were laid inside the garden, three of which were around its perimeter—one on the North side, the second on the South side, and the third opposite the circular road that formed the western boundary of the Capitol grounds. All three measured 20 feet in width and were bordered with beds 26 feet wide. The fourth walkway, 15 feet wide, was laid through the center of the garden and led to the elliptical pond.

These improvements did not, however, meet with universal approval. A complaint made to the Institute by the Commissioner of Public Buildings on June 9, 1827, reflects several of the problems deriving from the Institute's improvements. The Commissioner contended that the improvements to the garden were visually incompatible with the new section of the Washington Canal that had recently been "laid out along a line drawn through the middle of the Capitol and of the Mall. The foot-way, canals & plantation in the garden," he felt, "did not coincide with the line, but diverge from it at an acute angle. This discrepancy [was] so glaring and so offensive to the eye" that the Commissioner was convinced "every person visiting the Capitol would be grateful for its removal."

The placement of the garden at its present site, he explained, was made "with the expectation that it would become an ornamental appendage to the Capitol, and that under the eye of Congress they would be induced to foster it." Whether it became such an "ornament or deformity" largely depended upon the plan the Institute pursued in improving the site. Subsequently, the "deformity" cited by the Commissioner was found to be far less serious than originally thought and was quickly corrected.[13]

Plant Collection and Distribution

Meanwhile, the Institute embarked on a far-reaching effort to broaden its various collections. In March 1826, a three-member committee was appointed to meet with the heads of the various government departments and solicit their support in having our Nation's foreign representatives send the Institute "all subjects of natural history that may be deemed interesting." [14]

Three months later, a committee was appointed to prepare a report on collecting and preserving animal, vegetable, and mineral specimens.[15] That September, Secretary of the Treasury Richard Rush, an active Institute member, prepared a circular letter emphasizing President John Quincy Adams' personal interest in seeing our representatives throughout the world assist in bringing to the United States any foreign tree or plant that might, with proper cultivation, become useful to the American people. Accompanying Rush's letter was "Directions for Putting Up and Transmitting Seeds and Plants," prepared by Dr. James M. Staughton, a longtime member of the Institute.[16]

Both documents were sent to: (1) each Member of Congress with a request for them to send copies to constituents who might aid the botanic garden; (2) the Secretary of State for transmittal to each diplomat and commercial agent of

---

[13] Rathbun, The Columbian Institute, pp. 44–45. See also Minutes of the Columbian Institute, July 2, 1827, p. 259; August 13, 1827, p. 261; and November 5, 1827, pp. 272–273.

[14] Minutes of the Columbian Institute, March 4, 1826, p. 206.

[15] Ibid., May 13, 1826, p. 218.

[16] A copy of Secretary Rush's letter and Staughton's instructions appear in Introduction of Foreign Plants and Seeds, National Intelligencer, November 17, 1827, p. 2. Both documents are reproduced in Appendix 1.

the United States at home and abroad; (3) the Secretary of the Treasury for transmittal to each custom house or station; (4) the Secretary of the Navy for transmittal to each ship in commission; and (5) the Postmaster General for transmittal to various ports in different parts of the Union.[17]

A month later, Alexander McWilliams and Dr. Staughton presented a report to the Institute entitled "Best Means of Preserving Animal, Vegetable, and Mineral Specimens." Following their presentation, Asbury Dickins, Secretary of the Institute, drafted a letter to accompany the report. It was the Institute's hope, Dickins stressed, that its botanic garden would in time be able to collect and distribute throughout the United States "not only improved varieties of fruits and esculent and ornamental plants, already cultivated in this country, as well as valuable trees and plants that have been found indigenous in [this country, but also] all the vegetable production of other portions of the world, which can be adapted to the climate, and are desirable either for use or ornament." [18]

Secretary of the Treasury Rush's letter and the "Directions for Putting Up and Transmitting Seeds" were subsequently reprinted by Washington's *Daily National Intelligencer* on November 17, 1827, in the hope that their objective "might be better prompted." [19] A week later, the *Intelligencer* published Dickins' letter and the McWilliams-Staughton report without comment.[20] Then on November 30, the *Intelligencer* ran a lengthy story on the Institute, placing particular emphasis on the Garden's importance as well as the Administration's and Congress' enthusiasm for the project. "Let our Naval Officers," the *Intelligencer* urged,

(as recommended by the present patriotic Administration) bring home from their long cruises in the Mediterranean, Pacific, &c. the seeds of every plant indigenous in those countries, but strangers to ours, and present

---

[17] Minutes of the Columbian Institute, January 21, 1828.

[18] Letter From the Columbian Institute, National Intelligencer, November 24, 1827, p. 2. A copy of the McWilliams-Staughton Report is found in Minutes of the Columbian Institute, October 1, 1827, pp. 264–269. Dickins' letter appears in Ibid., November 5, 1827, pp. 270–272.

[19] Introduction of Foreign Plants and Seeds, National Intelligencer, November 17, 1827, p. 2.

[20] Letter From the Columbian Institute, p. 2.

them to the Columbian Institute, to be propagated and cultivated in their Botanic Garden; for which the ground is now properly prepared.

The climate of Washington, is also, it is believed, excellently adapted to naturalize almost every foreign plant that is capable of being brought to maturity in this country; and from thence they may be transferred to every part of the nation.[21]

Through the efforts of Secretary Rush and the *Intelligencer*, the objectives of the Institute quickly became well publicized and widely known. During the next several months, it received several letters accompanied by plants and seeds in response to Rush's letter and Staughton's collection instructions. Those responses, arranged in chronological order, are as follows:

1. *October 20, 1827.* F. Leandro de Sacranto of Brazil sent the following species: *Artocarpus incisa, Araucaria imbricata, Myrtus lucida, Myrtus jaboticaba, Myrtus lambuca, Annona squamosa, Eugenia jambos,* and *Petroa exelubilis.*[22]
2. *November 23, 1827.* Mr. J. L. Smith of Pennsylvania sent seeds of the *Litchi (Litchi chinensis)*, which he said were personally collected in China.
3. *December 14, 1827.* The records include only the first page of this letter; thus, the source is unknown. The writer stated that he had forwarded the Institute's circular to Charles L. Bonoparte, Prince of Marigana. As a result, the Prince sent the Institute grains of *Oryza mutica*, a rice of which, according to the letter, the "Chinese are jealous."
4. *January 6, 1828.* Mr. J. L. Cry sent seeds of Quachita lettuce collected near the Quachita River.

---

[21] The Columbian Institute, National Intelligencer, November 30, 1827, p. 2.

[22] *Artocarpus incisa* (syn: *A. altilis*) is commonly known as breadfruit; *Araucaria imbricata* (syn: *A. araucana*) is known as the monkey-puzzle tree; the three *Myrtus* species appear to be of no current botanical standing. They are probably misnamed species or cultivars of myrtle. *Annona squamosa* is commonly called sugar or custard apple. *Eugenia jambos* (syn: *Syzygium jambos*) is commonly called rose apple or malabar plum. *Petroa exelubilis* is a name of no current botanical standing.

5. *May 19, 1828.* Mr. Alex Roper sent plant material described in his letter as growing wild on the prairies of Montgomery County. He referred to one of the plants as the toothache tree and another as red root. The Latin names are not given. However, judging from Mr. Roper's description, the former is probably *Zanthoxylum americanum.* The latter could be one of several species commonly known as red root. This letter is particularly interesting, since reference is made to the medical value of the plants donated.
6. *Undated.* The records include only a portion of this letter which accompanied the following donations: "100 oranges with scions, 12 cuttings of figs, 9 palmetto roots, dates with seeds within, 150 palmetto seeds, 50 cuttings of grape vines, barley, flax, and wheat seeds." The letter stated that the palmettos and dates were intended as an experiment for the southern states. Also of interest is the following note written beneath the list: "12 of the grape vines directed to be sent to James Monroe, Esq." [23]

Numerous other letters are included among the Institute's records, but plant names are not given. The preceding probably represents only a small portion of the donations given to the Botanic Garden. The Institute's records also refer to plant material received prior to the 1827 circular, but unfortunately the specific names of these are seldom mentioned.

Although the Institute's records of the Botanic Garden's role in plant distribution have not survived, the following notice from the *Intelligencer* of May 22, 1828, clearly shows that this was one of the Institute's primary goals.

The Columbian Institute has just received from Tangier, in Morocco, some *Wheat* and *Barley* which, it is supposed, may form a useful addition to the stock of those grains already in the United States, particularly in the States and Territories South and Southwest of Washington. The Institute has also received some seeds and fruit of the date, which have been sent under a

---

[23] These letters are found in the Peter Force Papers, Series 8D, Item 24, Manuscript Division, Library of Congress.

belief that they may be successfully cultivated in the Southern parts of the Union. Tangier, whence these grains and seeds are brought, is in the latitude of 35 North; though black frosts are rare, white frosts are frequent there in January, February and March.

Those members of Congress who may desire to obtain a portion of either or all of these objects will please make known their wishes to Mr. Dickins, the Secretary of the Institute.[24]

Plants Cultivated in the Institute's Garden

The plants collected as a result of the widely distributed circular and accompanying instructions were presumably placed in the Botanic Garden. These plants, however, constituted only a portion of the material cultivated in the Garden during the Institute's history. In 1824, William Elliot, a member of both the Institute and the Botanical Society of Washington, prepared a complete "List of Plants in the Botanic Garden of the Columbian Institute," but did not identify the source of these materials or list the day they were acquired.[25]

Maintaining the Garden

Maintenance of the Garden itself was at best sporadic. The Institute never retained a full-time gardener and frequently employed no one in this capacity at all. As a consequence, when work was done on the Garden, it was accomplished by temporary help or by individuals who were occasionally allowed to occupy the house located on the grounds and farm a small portion of the land. There were a few instances when small sums of money were expended on cultivating, plowing, or fertilizing portions of the ground, but the total sum expended for such work was quite small.

John Foy, who during the existence of the Institute was the gardener for the Capitol grounds, periodically assisted in a supervisory capacity, but this was done in his spare time

---

[24] Editorial, National Intelligencer, May 22, 1828, p. 3. See also Pinkett, Early Agricultural Societies in D.C., p. 43.

[25] William Elliot's handwritten list is illegible in several places, as is indicated in the alphabetized version of the list reproduced in Appendix 2.

since the maintenance of the Botanic Garden was not part of his official responsibilities.[26]

OTHER ACTIVITIES OF THE INSTITUTE

Beginnings of a National Herbarium

Although the formation of a herbarium was not one of the original objectives of the Columbian Institute, herbarium specimens were received and stored at the Garden.[27] Existing records show that two major contributions of herbarium specimens were made. In 1820, Dr. William Darlington contributed specimens of American plants. Six years later, Dr. Alexander McWilliams donated specimens of plants native to the District of Columbia. Dr. McWilliams' contributions were, according to the records, arranged by the Linnean classification system. In neither instance were the number of plants or the species recorded.[28]

Presentation of Scholarly Papers

During the Institute's existence, twenty-six different members formally presented papers. William Lambert, the principal contributor, authored more than half of the total of eighty-five papers delivered. Lambert's forty-four presentations focused primarily on astronomical and mathematical subjects. His findings "evoked considerable response, for he discussed the importance to the country of determining Washington's prime meridian in order to free American chart makers and navigators from their dependence upon Greenwich, England."[29] It was mainly through his efforts that the

---

[26] In December 1821, the Institute authorized Foy to place "certain seeds and plants" in the Garden "during the pleasure of the Institute." Minutes of the Columbian Institute, December 1, 1821, p. 98. Four years later, Foy told the Institute "that the trees in the botanic garden are suffering for want of attention; and that if the Institute would employ a laborer," [he] "would supervise his work, without any charge." Shortly thereafter, a laborer was "employed for one month to work in the Botanic garden under the direction of Mr. Foy." Ibid., May 7, 1825, pp. 170–171.

[27] The fact that a national herbarium was not among the stated objectives of the Institute is said to account, at least in part, for the establishment of the Botanical Society of Washington in 1817.

[28] Rathbun, The Columbian Institute, pp. 55–57.

[29] Green, Washington: Village and Capitol, 1800–1878, p. 69.

creation of the National Observatory in Washington was recommended by the Institute and ultimately became a reality.

Other papers presented before the Institute covered such varied topics as botany, physics, mechanics, and the currency. It was intended that they be published in a single volume, but there never were sufficient funds to realize that goal.[30]

Collection of Objects

The Columbian Institute also actively sought contributions for its museum. In accordance with its objectives, the Institute was able to acquire a collection of mineral, zoological, and archeological specimens, as well as several other possessions of historical interest. Among the Institute's most interesting acquisitions were the regimentals worn by George Washington while Commander-in-Chief of the American Army during the Revolutionary War. Next to the Botanic Garden, however, the Institute's mineral collection was its most important asset. It included ores, rocks, and several different kinds of building stones.[31]

REASONS FOR THE INSTITUTE'S FAILURE

The Columbian Institute for the Promotion of Arts and Sciences disbanded in 1837 and formally went out of existence four years later. The limited success and short life span of the Institute can be attributed, in large measure, to its lack of financial support. It was forced to rely entirely on the revenue collected through membership dues and other small, rather insignificant, contributions.

As one of Washington's first learned societies, the Institute attracted some of the Nation's most prominent citizens. Numbered among its membership were three incumbent Presidents: James Monroe, who accepted the title of Patron of the Institute; and John Quincy Adams and Andrew Jackson, who were resident members while in the White House. Former Presidents John Adams, Thomas Jefferson, and James Madison were extended honorary membership.

---

[30] Rathbun, The Columbian Institute, p. 36.
[31] Ibid., pp. 54–58.

The Institute's membership at various times also included Vice President John C. Calhoun; the Marquis de Lafayette; fifteen United States Senators; twenty-eight Members of the House of Representatives; thirteen Cabinet secretaries; Benjamin Henry Latrobe, Charles Bulfinch, and George Hadfield, all of whom were involved in the design or construction of the Capitol; as well as James Hoban and Robert Mills, architects of the White House and the Washington Monument, respectively. In addition, there were distinguished representatives from the architectural and medical professions, business community, clergy, judiciary, local politics, the Army, and the Navy.[32]

Despite this glowing list of Washington luminaries, financial support from Congress was almost nonexistent.[33] A room was assigned for the Institute's use in the Capitol, but no other assistance for the organization was ever extended.[34] Also, the Institute's membership remained quite small throughout its twenty-one year existence, and it never was able to attract enough members to support its aims. The proceeds from dues were scarcely sufficient for incidental expenses. Numerous plans for raising additional funds, such as holding a lottery, selling public lots, soliciting contributions from citizens, and obtaining congressional support, were devised, but none was ever carried out.[35]

---

[32] A discussion of the Institute's membership is found in Rathbun, The Columbian Institute, pp. 18–23; and Pinkett, Early Agricultural Societies in D.C., p. 40.

[33] One particularly unfortunate occurrence took place in this regard in 1837 when John McArann of Philadelphia offered to sell the Institute a collection of exotic and indigenous plants including a coffee tree, night-blooming cereus, sago and palm trees, sugar cane, cinnamon tree, and a great variety of botanically interesting specimens. Although McArann had the written support of forty members of the Horticultural Society of Pennsylvania as well as the support of the Institute's membership, Congress refused to provide the necessary funds for the transaction to be completed. U.S. Congress, House, Committee on Public Buildings, House Report No. 290, 24th Cong., 2d Sess. Washington: Blair & Rives Printers, 1837, 5 pp. (Serial No. 306).

[34] Wilhelmus Bogart Bryan, A History of the National Capitol, 2 vols., New York: Macmillan Company, 1916, v. 2, p. 30.

[35] Pinkett, Early Agricultural Societies in D.C., p. 41.

Not only did the Columbian Institute want for an adequate financial base to carry out its most basic aspirations; it also suffered from a lack of leadership. Although there was considerable enthusiasm for the Institute during its first decade, it essentially was an organization composed of gentlemen who were usually occupied with official or professional duties. As a consequence, for the most part, it was run by amateurs.

Had the Institute possessed adequate funds, it could have employed experts in a number of different fields. Unfortunately, in most instances its members were willing to give the Institute little more than their names. The Institute's meetings were poorly attended and volunteers for service few. Given these realities, the Institute's objectives could not possibly have been realized.

In 1841, the Columbian Institute merged with the Historical Society of Washington, which had been founded five years earlier, and the objects possessed by the Institute were transferred to the National Institute for the Promotion of Science, a predecessor to the Smithsonian Institution.[36] Two decades later, these objects were turned over to the Smithsonian. The Institute's legacy was that it provided a foundation framework for the United States Botanic Garden as well as providing the objects that would ultimately help form the basis of the initial Smithsonian collection.

With the demise of the Columbian Institute the site of the Botanic Garden reverted to the Federal Government. Not until 1850 did the Government again express an interest in the site as it sought a new location for the botanical collection of Lieutenant Charles Wilkes' Exploring Expedition of 1838–1842.[37]

---

[36] Second Bulletin of the Proceedings of the National Institute for the Promotion of Science, Washington: Printed by Peter Force, 1842, pp. 94, 113.

[37] Ibid., p. 43

Lieutenant Charles Wilkes
Naval Photographic Center, Washington, D.C.

CHAPTER II

THE WILKES EXPEDITION (1838–1842): A SECOND CHANCE FOR THE NATIONAL BOTANIC GARDEN

Although the Columbian Institute failed to establish a permanent national botanic garden, it did succeed "in establishing a precedent for creating such an institution under the auspices of the Federal Government."[1] Five years after the Institute's demise in 1837, the idea of a botanic garden was revived, but the impetus for that moment began much earlier. On the morning of August 18, 1838, Lieutenant Charles Wilkes led a squadron of six ships and 440 men from Hampton Roads, Virginia, on a voyage that was to figure prominently in the development of a permanent botanic garden. During their ensuing four-year, 87,000-mile voyage, Wilkes' Expedition would circumnavigate the globe, complete extensive surveys of the Pacific Ocean, and confirm the existence of the continent Antarctica. Altogether, this mission would cost the United States Government $928,183.62.[2]

What prompted a struggling young nation to undertake such a major project? A synopsis of the Expedition, delivered by Wilkes before the National Institute (the successor to the Columbian Institute) shortly after the conclusion of the Expedition in June 1842, provides several important answers

---

[1] Karen D. Solit, The U.S. Botanic Garden, American Horticulturist v. 61, April 1982, p. 5.

[2] Daniel C. Haskell, The United States Exploring Expedition 1838–1842, New York: New York Public Library, 1942, p. 6. For background on the Wilkes Expedition see G. S. Bryan, The Wilkes Exploring Expedition, United States Naval Institute Proceedings, v. 65, October 1939, pp. 1452–1464; Daniel Henderson, The Hidden Coast: A Biography of Admiral Charles Wilkes, New York: William Sloan Associates Publishers, 1951, pp. 29–202; William James Morgan, David Tyler, Joye L. Leonhart, and Mary F. Loughlin, eds., Autobiography of Rear Admiral Charles Wilkes, U.S. Navy 1798–1877, Washington: Naval History Division, Department of Navy, 1978, pp. 321–548; Robert E. Morsberger, The Wilkes' Expedition: 1838–1842, American History Illustrated, v. 7, June 1972, pp. 4–10, 45–49; and Herman J. Viola and Carolyn Margolis, ed., Magnificent Voyagers: The U.S. Exploring Expedition, 1838–1842, Washington: Smithsonian Institution Press, 1985.

to that question. Wilkes saw the primary objective of the United States Exploring Expedition of 1838–1842 as being the "promotion of the great interest of commerce and navigation."[3] Others, in more precise terms, argue that the "whaling industry, then in its heyday, appears to have supplied the principal motive" for the Expedition.[4] It was hoped that the Wilkes voyage would provide more detailed information regarding remote and poorly charted regions and allow the United States an opportunity to compete with the great nations of Europe in geographic exploration.

Every opportunity was also taken by the Expedition, when it was "not incompatible with the great purpose of the undertaking, to extend the bounds of science and to promote the acquisition of knowledge."[5] The collection of plant material was neither the prime objective of the Wilkes Expedition nor its principal accomplishment, but the botanic specimens it acquired were extremely significant and ultimately formed the nucleus of the collection housed in the Botanic Garden's first conservatory.[6]

---

[3] Charles Wilkes, Synopsis of the Cruise of the U.S. Exploring Expedition During the Years 1838, '39, '40, '41, & '42, Washington: Peter Force, 1842, p. 6.

[4] Bryan, The Wilkes Expedition, p. 1452. See also G. S. Bryan, The Purpose, Equipment and Personnel of the Wilkes Expedition, Proceedings of the American Philosophical Society, v. 28, June 29, 1940, pp. 551-552; and Morsberger, The Wilkes Expedition, p. 8.

[5] Wilkes, Synopsis of the Cruise, p. 6. See also Harley Harris Bartlett, The Reports of the Wilkes Expedition and the Work of Specialists in Science, Proceedings of the American Philosophical Society, v. 82, June 29, 1940, pp. 601–705; and Edwin G. Conklin, Connection of the American Philosophical Society with Our First National Exploring Expedition, Ibid., p. 519–541; and Louis N. Feipel, The Wilkes Expedition: Its Progress Through Half a Century: 1826–1876, United States Naval Institute Proceedings, v. 40, September–October 1914, pp. 1323–1350.

[6] John Hendley Barnhart, Brackenridge and His Book on Ferns, Journal of the New York Botanical Garden, v. 20, June 1919, pp. 117–124. It is interesting to note that Charles Wilkes became a member of the Columbian Institute in 1833, five years before the Expedition set sail. The Institute was asked by Secretary of the Navy Samuel Southard to offer suggestions or views as to the scientists that should be sent, instruments needed, subjects of scientific inquiry, and the like. And a committee of the Institute, appointed to investigate these subjects, reported their findings to Secretary Southard.

EXPEDITION'S SCIENTIFIC CORPS

To assure the successful attainment of scientific data, several prominent scientists were selected to accompany the Expedition, among whom were William Rich, a botanist; William D. Brackenridge, a horticulturist; and Charles Pickering, a naturalist. As the Expedition unfolded, William Rich, as Captain Wilkes soon discovered, had a "gentlemanly and quiet demeanor," but was "illy qualified" for the task at hand.[7] Conversely, William Brackenridge, while not well versed in the technical forms of descriptive plant taxonomy or Latin, was considered an excellent field botanist. Among other accomplishments, he is credited with discovering the California pitcher plant, *Darlingtonia californica*, the "most notable American plant found by the Expedition." [8]

ANYTHING BUT SMOOTH SAILING

The Wilkes Expedition was one of the most far-reaching terrestrial explorations in our Nation's history and certainly one of its most important. It was not, however, devoid of problems. Charles Wilkes was under no illusions as to the condition of the ships assigned him. All of them were "ill-equipped for making such a voyage and their crews were quite restive." [9] Before the Expedition was over, two ships were lost: one during the first year off Cape Horn with its

---

[7] Morgan, Autobiography of Rear Admiral Charles Wilkes, p. 382. For background on the members of the Scientific Corps see Ibid., pp. 381-382.

[8] Bartlett, Reports of the Wilkes Expedition, p. 682. At the outset, renowned botanist Asa Gray was chosen as head botanist for the Expedition, but he was to resign before the Expedition ever got under way. Instead, during the lengthy delay in the Expedition's departure following his selection, Gray accepted an appointment as Professor of Natural History at the newly chartered University of Michigan. For biographical information on Brackenridge see Bartlett, Reports of the Wilkes Expedition, pp. 675-676.

[9] Bryan, The Wilkes Expedition, p. 1453. See also Bryan, Purpose, Equipment and Personnel of the Wilkes Expedition, pp. 554-556; James D. Hill, Charles Wilkes—Turbulent Scholar of the Old Navy, United States Naval Institute Proceedings, v. 57, July 1931, p. 868; Morgan, Autobiography of Charles Wilkes, pp. 385-388; and W. Patrick Strauss, Preparing the Wilkes Expedition: A Study in Disorganization, Pacific Historical Review, v. 28, August 1959, pp. 221-232.

seventeen-man crew aboard; the other, without loss of life, in 1841 in the Columbia River. Another, considered to be unseaworthy, was sold during the voyage. Also, during the lengthy voyage two officers were killed by Fijian natives. These misfortunes, coupled with illness, perilous weather, and dissatisfied crews, made for an eventful and dramatic journey.[10]

Wilkes' complex personality reportedly resulted in several other problems as well. Many at the time considered him to be a highly intelligent scientist and a dedicated officer. Others thought Wilkes to be rash, impetuous, hot tempered, and a harsh disciplinarian. These personal traits contrast sharply with those of Wilkes' aunt, Mrs. Seaton, later to become Mother Seaton, with whom he lived for a short time while a boy. Mother Seaton was canonized in 1974—the first American saint. Often he was at odds with both his superiors at home and his subordinates at sea. He was remembered by some as a snob, irascible, self-righteous, and opinionated. Understandably, his relationship with the Expedition's scientists was far from harmonious.[11]

EXPEDITION'S ACCOMPLISHMENTS

Despite these and numerous other problems, the accomplishments of the Expedition were considerable, especially in light of its being the first such venture undertaken by the United States. Wilkes and his crew surveyed vast portions of the Pacific Ocean. They also explored many of the islands in the South Pacific, as well as Australia, New Zealand, the Hawaiian Islands, the Philippines, Singapore, the Cape of Good Hope, St. Helena, and the coast of the Pacific Northwest.

Considerable time was also spent cruising approximately 1,600 miles of the Antarctic coastline. This latter venture confirmed that Antarctica was a continent—a continuous land mass, rather than a series of islands. This was the

---

[10] Bryan, The Wilkes Expedition, pp. 1455, 1457, 1460–1461.
[11] Hill, Charles Wilkes—Turbulent Scholar of the Old Navy, pp. 868–869; and Morgan, Autobiography of Charles Wilkes, pp. 383–385.

voyage's most notable accomplishment and the one on which its reputation rests.¹²

Upon the Expedition's arrival at a pre-selected port of call, an astronomical station was usually set up and a survey begun, and the Expedition's scientists conducted on-site investigations of geographic, botanic, and other ecological conditions. Approximately 10,000 plant species were studied during these investigations, and three to five herbarium specimens of each were collected. In addition, propagation material from an untold number of plants was also gathered.

The Expedition's bounty included fossils, minerals, shells, insects, animals, and marine life along with seeds, nuts, living plants, and dry specimens. Brought back as well were numerous drawings prepared by the two artists who traveled with the Expedition, a number of artifacts of significant historical value, and more than 180 charts containing descriptions of landmarks, harbors, and other points of interest prepared during the voyage.¹³

EXPEDITION'S RECORD: A PERMANENT LEGACY

There also exists in the Library of Congress and Naval Research Library in Washington a permanent published record of the investigation and exploration of the four-year Wilkes Expedition. It fills twenty-four sizeable volumes. The first five volumes comprise a narrative of the voyage written by Charles Wilkes, including several descriptions and pictures of native flora.¹⁴

---

[12] Bryan, The Wilkes Exploring Expedition, p. 1452-1464; Mark Cooley, The Exploring Expedition in the Pacific, Proceedings of the American Philosophical Society, v. 82, June 29, 1940, pp. 707-719; Henderson, Hidden Coasts, pp. 46-200; William Herbert Hobbs, The Discovery of Wilkes Land, Antarctic, Proceedings of the American Philosophical Society, v. 82, June 29, 1940, pp. 561-582; and Morsberger, The Wilkes' Expedition, pp. 6-10, 45-49.

[13] Feipel, The Wilkes Exploring Expedition, p. 1346; and Robert Park MacHatton, Heritage of the Navy, United States Naval Institute Proceedings, v. 68, July 1942, p. 967.

[14] Charles Wilkes, Narrative of the United States Exploring Expedition During the Years 1838, 1839, 1840, 1841, 1842, Philadelphia: Lea & Blanchard, 1845. For background on how the Expedition's records came to be published see Bartlett, Reports of the Wilkes Expedition, pp. 630-635; Henderson, The Hidden Coast, pp. 214-218; and Morgan, Autobiography of Charles Wilkes, pp. 535-548.

Other volumes of particular interest to the development of the U.S. Botanic Garden are volumes 15 and 18, which discuss the *Phanerogamia* (seeds and flowering plants) collected by the Expedition;[15] volume 16, which is devoted to the ferns that were studied;[16] and volume 17, which includes descriptions of the lichens, fungi, mosses, and algae Wilkes encountered.[17] A portion of volume 17 is also devoted to the seed plants collected in the Pacific Northwest.

The Wilkes Expedition is interesting from a literary point of view as well. The determined and domineering Wilkes is said to have served as a model for Captain Ahab in Herman Melville's classic *Moby-Dick*.[18]

AMERICA'S RESPONSE TO THE EXPEDITION

Despite the importance of the Expedition, Wilkes and his men received far from a hero's welcome upon their return. The reasons for this reaction are complex and reflected reac-

---

[15] Volume 15 was authored by botanist Asa Gray. In a June 1848 letter to John Torrey, a well known botanist of the period, Gray explained that he accepted the task on the understanding that part of the required research could be performed in one of the established herbariums abroad. The work involved describing, and in many cases identifying, the dried plant specimens collected during the Expedition. Gray felt this task could only be accomplished by studying the collections abroad, the quality of which were not equaled in the United States. He predicted that the work would take approximately five years to complete. Wilkes at once accepted his terms. Jane Loring Gray, ed., Letters of Asa Gray, New York: Lenox Hill Pub. & Dist. Co., 1893, p. 359. Gray actually completed volume 15 in 1854. Volume 18, also written by Gray, was, for a variety of reasons, never published. The manuscript is held by the Gray Herbarium at Harvard University. For background on Asa Gray see footnote 45. See also Bartlett, Reports of the Wilkes Expedition, pp. 664–673.

[16] Volume 16, entitled Filices, was written by James Brackenridge. A summary of this volume is found in Bartlett, Reports of the Wilkes Expedition, pp. 673–679.

[17] Volume 17 was edited by Asa Gray. Four sections of the volume were written by four different authors, none of whom actually were members of the Wilkes Expedition. The fifth, devoted to the seed collections, was prepared by John Torrey. For a summary of this volume see Ibid., pp. 679–682.

[18] David Jaffe, Literary Detective Harpoons a Whale of a Tale, Potomac Magazine (Washington Post), June 2, 1963, pp. 18–19; and David Jaffe, The Stormy Petrel and the Whale: Some Origins of Moby-Dick, Washington: University Press of America, 1982, pp. 7–38.

tions to Wilkes' behavior during the voyage, scientific disputes arising from the Expedition, and the fact that the Van Buren Administration, which had sponsored the venture, had been turned out of office by the Whigs the previous November.

Charles Wilkes was vilified upon his return. In 1842 he was court-martialed for conduct unbecoming an officer as a result of his mistreatment of the Expedition's crew. Although Wilkes was subsequently acquitted, his reputation and that of the Expedition had been severely tainted. He was criticized as well by various explorers of the day, including Englishman James Clark Ross, who claimed that Wilkes' determinations regarding the Antarctic land mass were incorrect. At the time, Ross' opinion was widely accepted. Today, however, it is widely agreed that Wilkes did in fact establish the existence of a solid Antarctic land mass, part of which was later named for him.[19] Acknowledged as well is the significant contribution of the Wilkes Expedition to the birth of a permanent U.S. Botanic Garden.

DISPOSITION OF THE PLANT MATERIAL COLLECTED

By the time Wilkes returned to the United States and entered New York harbor on June 9, 1842, he had covered 87,000 miles. The returning ships carried over 10,000 plant specimens and seeds, and over 250 live plants.[20]

Initially, the Expedition's collection of plants was placed in a lot located behind the Old Patent Office in Washington. By the fall of 1842, Congress had approved funds for a greenhouse to protect the plants, and William D. Brackenridge had assumed responsibility for their care.

That November, Brackenridge, in a report to Curator Charles Pickering of the National Institute, announced the

---

[19] Henderson, The Hidden Coast, pp. 119–124; Hill, Charles Wilkes, pp. 869–870; Morgan, Autobiography of Charles Wilkes, p. 520–525; and Morsberger, The Wilkes Expedition: 1838-1842, p. 48.

[20] Goode, The Genesis of the National Museum, pp. 353–354.

completion of a 50-foot-long greenhouse that would house the plant collection at the Patent Office. Because of the numerous plants the Institute had acquired from the Wilkes Expedition, Brackenridge announced, the Institute now possessed one of the "most extensive and varied botanical collections." More than 500 species were already in cultivation and another 1,100 plants had been placed in pots.[21]

Brackenridge's report also contained the following list of the live plants and herbarium specimens that had been collected at the various places visited by the Expedition. These included:

Maleira	300	Tongatabu	236
Cape de Verde Islands	60	Fiji Islands	786
Brazil	989	Low Coral Islands	27
Patagonia (Rio Negro)	150	Sandwich Islands	883
		Oregon Country	1,218
Terra del Fuego	220	California	519
Chile and Chilean Andes	442	Manila	381
Peru and Peruvian Andes	820	Singapore	80
		Mindanao	102
Tahiti	288	Tulu Islands	58
Samoa, or Navigator Islands	457	Mangsi Islands	80
		Cape of Good Hope	330
New Holland	789	St. Helena	20
New Zealand	398		
Lord Auckland Island	50	Total number of species	[22] 9,674

---

[21] Ibid., pp. 353–354; and Rathbun, The Columbian Institute, p. 51. See also Bartlett, Reports of the Wilkes Expedition, pp. 676–677.

[22] Goode, The Genesis of the National Museum, p. 353. The items in the published list add to 9,683. This discrepancy is not explained.

Brackenridge reported,

> The number of seeds brought and sent home by the Expedition amounted to 684 species most of which have been sent all over the country. Several cases of live plants were also sent home, of the existence of which there are no traces. The live plants brought home by the squadron amounted to 254 species, and these now form part of the greenhouse collection.[23]

Given the increasing number of plants the Institute was receiving in exchange for the Wilkes Expedition seeds it was distributing, Brackenridge felt that within a year the botanical collection would outgrow the new greenhouse.[24] Brackenridge's projections proved to be correct; subsequently, two additions to the Patent Office greenhouse were constructed under the direction of the Joint Committee on the Library, and a small growing area was made available.[25]

The Wilkes collection was placed under the "direction and control of the Joint Committee on the Library" on March 3, 1843, the rationale being that the Committee had, some months earlier, been given the responsibility of supervising the publication of the results of the Expedition.[26] At the same time, the Committee officially "appointed and directed" Brackenridge to take "charge of and preserve" the Expedition's botanical and horticultural specimens.[27]

At the outset, all of the Expedition's collections were placed in the custody of the National Institute. Then in July 1843, the Joint Committee on the Library appointed the Commissioner of Patents as custodian of all Government

---

[23] Goode, the Genesis of the National Museum, pp. 353–354. See also Bartlett, Reports of the Wilkes Expedition, p. 678; and Rathbun, The Columbian Institute, p. 51.

[24] Goode, the Genesis of the National Museum, p. 354.

[25] 5 Stat. 642, 691. A comprehensive explanation of the expansions to the Patent Office greenhouse is found in Rathbun, The Columbian Institute, pp. 51–52.

[26] 5 Stat. 642; and U.S. Congress, Senate, Estimates of Expenditures on the Botanic Garden, 1850–1907, Senate Document No. 494, 60th Cong., 1st Sess., Washington: U.S. Govt. Print. Off., 1908, p. 4.

[27] U.S. Congress, Joint Committee on the Library, The Botanic Garden and Its Relation to the Joint Committee on the Library, July 1, 1912, Washington: U.S. Govt. Print. Off., 1912, pp. 3–4.

collections housed in the Patent Office, including those gathered by the Expedition. In August 1843, Captain Wilkes was placed in charge of the artifacts collected by the Expedition, which were placed on public display in the upper hall of the Patent Office. Brackenridge remained in charge of the botanical specimens.[28]

Many of the plants collected by the Wilkes Expedition, however, were never included within the greenhouse collection. Many of the plants that had been sent home during the course of the Expedition, as Brackenridge noted in his report, were either stolen or lost at sea. Others arrived in poor and unrevivable condition. In several instances, shipments of plants and other items were first sent to the Peale Museum, in Philadelphia, before being forwarded to Washington.[29]

As a consequence, it is reasonable to assume that most of the surviving plants arrived with the Expedition in 1842. These plants, as Wilkes explains in his *Autobiography*, "were transported in Loddiges cases which afforded full protection to them whilst passing through the various zones we necessarily had to go through."[30]

Even after the plants had actually arrived in Washington, there were still numerous challenges and annoyances associated with their preservation. "No sooner had it become known that the Government had a greenhouse," Wilkes wrote, than requests for flowers and plants became of such a magnitude that if they had "been acceded to, in a very short time we should not have had a plant remaining."[31]

To obviate this problem, Wilkes instructed Brackenridge to refuse all requests for flowers and plants and to use his name as rationale. "In several instances Senators and Members of Congress," according to Wilkes, "became quite

---

[28] Ibid. See also MacHatton, Heritage of the Navy, p. 968; and Morgan, Autobiography of Charles Wilkes, pp. 527–529.

[29] "After a while," as Wilkes points out in his *Autobiography*, "there were plants discovered in the United States which were known" to have been sent home by the Expedition. But in most instances he was unable to convince the individuals who had obtained them to return them. Morgan, Autobiography of Charles Wilkes, p. 530.

[30] Ibid., p. 582.

[31] Ibid., p. 529.

The large specimen of *Angiopteris evecta* (the Vessel Fern), now housed in the U.S. Botanic Garden Conservatory, was presumably propagated from a specimen brought to Washington, D.C., by the Wilkes Expedition.

irritated over this policy" and "swore they would make no appropriation for the care of the collection." Even the President's wife, Julia Tyler, paid "him a visit and demanded both plants & flowers. She was met by a positive denial and left in a huff." [32]

---

[32] Ibid., pp. 529–530.

An entirely different perspective of the Patent Office greenhouse can be gleaned from a 1844 article by C.M. Hovey, editor of the *Magazine of Horticulture*. Hovey found Brackenridge to be most accommodating and his garden most impressive. "Another season, under [Brackenridge's] attentive care," Hovey was convinced, would result in an even "better development of the habits and character of many of the more rare and tropical species." By that time, Brackenridge would also "have multiplied many of the plants, to such a degree, that they may, if such is the intent of government, be distributed among nurserymen."[33]

NEED FOR A NEW LOCATION

Five years later, in 1849, it became necessary to expand the Patent Office and find a new location for the botanical collections of the Wilkes Expedition. The site selected was at the east end of the Mall at the foot of Capitol Hill. Although many new species would be acquired to fill the new conservatory soon to be built there, the Wilkes Expedition contributions still formed the nucleus of the collection.[34]

Today, there are still two plants in the United States Botanic Garden collection that are considered to have been part of the Wilkes Expedition bounty. Both are described in the botanical journals of the Expedition. These are the *Angiopteris evecta*, the Vessel Fern; and the *Zizyphus jujuba*, the Chinese Jujube. A large specimen of the former species is housed in the Conservatory. The *Zizyphus* is located in the Frédéric Auguste Bartholdi Park, opposite the Conservatory, on Independence Avenue.

---

[33] C.M. Hovey, Experimental Garden of the National Institute, Magazine of Horticulture, v. 10, March 1844, p. 82. For his description of the garden, see Appendix 3.

[34] Solit, U.S. Botanic Garden, p. 5.

WILKES PLAQUE

The dedication ceremony of the bronze plaque honoring Rear Admiral Charles Wilkes occurred on November 12, 1985. This event took place in the United States Botanic Garden Conservatory Orangerie, where the plaque is installed.

Its inscription reads:

Admiral Charles Wilkes
April 3, 1798—February 8, 1877

This plaque is established in memory of Admiral Charles Wilkes, navigator and scientist, whose Pacific and Antarctic Expedition (1838–1842) yielded hundreds of exotic plants which formed the core collection of the United States Botanic Garden. The Garden was established in 1842 after Wilkes had successfully persuaded Congress to support scientific care and study of these specimens.

Admiral Wilkes recorded the results of this important expedition in his five-volume work, "Narrative of the United States Exploring Expedition." He discovered the Antarctic continent, where Wilkes Land bears his name. Known for his determination, dedication and strength of leadership, Charles Wilkes was one of the most colorful of the early American Naval Officers.

CHAPTER III

THE UNITED STATES BOTANIC GARDEN: NEW FACILITIES IN A FAMILIAR SETTING

By mid-May 1850, Congress had authorized the expenditure of $5,000 for the construction of a botanical conservatory on the Mall.[1] That fall, the plants housed in the Patent Office greenhouse were moved to the new structure, which had been built on the exact site previously occupied by the Columbian Institute's botanic garden.

Together the conservatory and accompanying grounds occupied ten acres extending from First to Third Streets between Pennsylvania and Maryland Avenues, Southwest. Although the new site "offered ample room for the care and preservation of the botanical collection," the funds Congress had appropriated for the project were not sufficient to provide for an adequate drainage system.[2] As a consequence, the new garden was plagued with many of the same problems that the Columbian Institute had faced. During the thirteen years since the Institute had used the site, it had become an abysmal swamp, subject to the ebb and flow of the Tiber Creek that traversed the property, and had been used as a dumping ground. All that remained of the original garden were two post oaks (*Quercus stellata*).

William D. Brackenridge continued to be in charge of the botanic collection at the new facilities and retained the title of horticulturist until the summer of 1854. Captain Charles Wilkes remained as supervisor of the new greenhouses until that fall.[3]

---

[1] 9 Stat. 427. The appropriation for the construction of the new greenhouses was expended under the direction of the Joint Committee on the Library.

[2] Rathbun, The Columbian Institute, pp. 52-53.

[3] Ibid., pp. 52-54. In 1855, William Brackenridge purchased thirty acres near Baltimore, where he lived until his death on February 3, 1893. For several years he was "horticultural editor of the *American Farmer*, but he spent most of his energies as a nurseryman and landscape architect." Allen Johnson and Dumas Malone, eds., Dictionary of American Biography, 20 vols., New York: Charles Scribner's Sons, 1928-1937, v. 2, p. 546. See also W.D. Brackenridge, Gardner's Age, v. 26, December 1884, p. 376; and W.D. Brackenridge, Meehan's Monthly, v. 3, March 3, 1893, p. 47.

In August 1856, in recognition of its increased stature, the Garden was officially named the United States Botanic Garden, its maintenance was for the first time specifically placed under the jurisdiction of the Joint Committee on the Library, and regular annual appropriations were begun. Although the Garden since 1843 had been under the "direction and control" of the Joint Committee, responsibility for its maintenance had rested with the Commissioner of Public Buildings.[4]

THE WILLIAM SMITH YEARS (1853–1912)

Meanwhile, in 1853, a twenty-two-year-old Scotsman, William R. Smith, on the recommendation of the eminent botanist Sir Joseph Hooker, was hired as a gardener and embarked on a career with the Garden that would span more than half a century. A decade later, Smith became the first superintendent of the Garden.

Smith brought considerable experience to his new position. While still a teenager, he was an assistant to the gardener for Lord Elcho at Haddington. Subsequently, he became a student at the Royal Botanic Gardens at Kew. After graduating from Kew Gardens in 1852, Smith left for Philadelphia, where he worked with the famous gardener Dudassie on Chestnut Street. A short time later, he received his appointment to the United States Botanic Garden.[5]

---

[4] 11 Stat. 104; U.S. Congress, House, Committee on the Library, United States Botanic Garden, Preliminary Report and Memorandum on the U.S. Botanic Garden and Kindred Institutions Together with Certain Recommendations Looking Through Its Improvement as a Scientific, Educational, and Aesthetic Accomplishment, Committee Print, 73d Cong., 2d Sess., Washington: U.S. Govt. Print. Off., 1934, p. 16; U.S. Congress, Joint Committee on the Library, The Botanic Garden and Its Relationship to the Joint Committee on the Library, Washington: U.S. Govt. Print. Off., 1912, pp. 6–7; and H.P. Caemmerer, Washington: The National Capitol, Washington: U.S. Govt. Print. Off., 1932, p. 186.

[5] Answers Last Call: Passing of William R. Smith at the Age of Eighty-Four, Washington Sunday Star, July 7, 1912, p. 2, pt. 1; Catherine Francis Cavanagh, A Great Champion of Burns and Masonry, New Age Magazine, v. 14, January 1911, p. 60; and James MacPherson Jarrett, William Robertson Smith—Founder of the Saint Andrew's Society of Washington, D.C., Newsletter of the Saint Andrew's Society of Washington, D.C., No. 123, July 1974, p. 4.

Soon after moving to Washington, Smith prepared the first comprehensive catalog of the Garden's plants.⁶ He was also kept busy collecting the Garden's highly sought-after seeds and distributing them around the country. One of the Garden's most widely distributed plants was the Boston Ivy.⁷

During Smith's nearly sixty years at the Botanic Garden, the grounds became in some respects his private estate. It was there, in the ivy-covered brick cottage nestled among the Garden's greenhouses and plantings, that he made his home and amassed in the cottage's 10-by-12-foot sitting room the world's foremost collection of the works of the poet Robert Burns.⁸

As Smith's collection continued to grow, it ultimately spread to a second room in the cottage. Although he personally had insufficient funds to buy the rare editions that came to his attention, he was able to add to his collection on a consistent basis because of numerous gifts from prom-

---

⁶ A list of the plants extracted from the catalog is reproduced in Appendix 5. In the introduction to the catalog Smith states "that the majority of the plants in this list are the results of the United States Exploring Expedition, commanded by Captain Wilkes with several additions by other officers of the navy and army. Mr. Brackenridge by a judicious system of exchanging has obtained many important additions. Several of the plants first discovered by the expedition are now to be found wherever an exotic collection exists." William R. Smith, A Catalog of Plants in the National Conservatories, A Popular Catalogue of the Extraordinary Curiosities in the National Institute Arranged in the Building Belong to the Patent Office, Washington: Alfred Hunter, 1854, p. 64.

⁷ Smith had a very special feeling for the Boston Ivy (or *Parthenocissus tricuspidata* [also once known as *Ampelopsis veitchii*], Boston Ivy's scientific name). It was Smith who first brought it to this country and as a consequence "was the foster father of the most popular vine in America." Subsequently, he nurtured the cuttings "set out on the south side of the brick building near the west end of the Botanic Garden in which [were located the] offices of the superintendent and assistant superintendent," and every year would send the seeds of this vine all over the country. Ancestral Vine of All American Ampelopsis in the Botanic Garden, Washington Sunday Star, October 20, 1907, p. 6, pt. 4.

⁸ Smith's collection comprised 5,000 volumes, including "700 copies of the 900 editions of Burns' poems, songs, and letters. Nearly 200 volumes were a duplicate of Burns' own library. Some 4,000 volumes were made up of Burns' biographies, eulogies, and quotations." Jarrett, William Robertson Smith, p. 5.

William R. Smith, first Superintendent of the Botanic Garden.
Scottish Rite Temple, Washington, D.C.

inent people throughout the world, such as Andrew Carnegie.[9]

Smith was also considered to be the "Father of the Society of American Florists and Ornamental Horticulturists," and was a member of the Association of Oldest Inhabitants of

---

[9] After Smith's death in July 1912, Carnegie assumed responsibility for the Burns collection and arranged for it to be cataloged by William Thompson of the Public Library of Edinburgh, Scotland. Today, the Smith collection is housed in the Burnsiana Room in the Masonic Temple of the Supreme Council Thirty-Third Degree in Washington. Ibid.

the District of Columbia. "His cottage in the peaceful quiet of the Botanical Gardens" served as a refuge for many a prominent Washingtonian. Members of both Houses of Congress strolled down from Capitol Hill not only to visit the Garden, but also to spend time with the scholarly Smith. While pointing out the Garden's botanical collections, Smith would often quote Burns or share insights into his latest acquisition of the late poet's work.[10]

He invited many of his prominent friends to plant memorial trees in the Botanic Garden. But because he feared that relic-hunting tourists might appropriate pieces of these cherished trees as souvenirs, Smith left them unlabeled.[11] Still, he took great pride in escorting interested visitors around the grounds and pointing out the living memorials. Unfortunately, none of those trees has survived to the present time.[12]

During his long tenure as superintendent of the Garden, Smith met many Congressmen who believed the facilities existed solely for their personal use. In each instance, Smith gently but firmly informed them of the Garden's limitations.

Once, while showing a group of Congressmen through the conservatories, he spoke lovingly of the rare orchids growing in one of the houses; whereupon one Member spoke up and asked: "What are these darned things worth anyhow?" Smith replied solemnly, "My dear sir, if the Great Architect of the Universe had been considering economy when he created *you*, he would have put you on four feet and fed you on grass."[13]

"No doubt that [the] Congressman left the gardens vowing vengeance" on the outspoken gardener, but nothing ever came of the incident. Despite his outspokenness, one writer in 1911 was prompted to suggest that the popular Smith was "as secure in his position as the Chief Justice of the United States."[14] His independent air even extended to an unwill-

---

[10] Ibid.; Cavanagh, A Great Champion of Burns and Masonry, p. 62; and Answers Last Call, Sunday Star, July 7, 1912, p. 2, pt. 1.
[11] Cavanagh, A Great Champion of Burns and Masonry, p. 62.
[12] Descriptions of a number of these species are included in Appendix 4.
[13] Cavanagh, A Great Champion of Burns and Masonry, p. 62.
[14] Ibid.

ingness to cooperate when Congress sought to raise his salary, saying that he was being paid quite sufficiently for his needs.[15]

THE GARDEN IN THE LATTER HALF OF THE NINETEENTH CENTURY

A description published in 1875, during Smith's tenure, tells us that the Garden was open from 9 a.m. to 6 p.m. There were "two main entrances for pedestrians, one opposite the main central West gate of Capitol Park and the other on Third St., opposite the east end of the Drive. Each entrance consisted of four marble and brick gate piers, with iron gates." The Garden comprised ten acres "surrounded by a low, brick wall, with coping and iron railing, and [was] laid out in walks, lawns and flower-beds." Wheeled vehicles were not permitted on the grounds.[16]

The Bartholdi Fountain

One of the major attractions of the Garden, located north of the Main Conservatory, was the Bartholdi Fountain. Originally, this thirty-foot-high sculpture, the work of Frenchman Frédéric Auguste Bartholdi, was designed for the Philadelphia Centennial Exposition of 1876, where it stood in front of the main entrance. Following the closing of the Centennial Exposition in 1877, the United States Government purchased Bartholdi's fountain for $6,000 and had it erected on the grounds of the Botanic Garden.[17]

When it was initially cast, the "fountain epitomized the artistic use of cast iron, which had previously been looked upon with disdain by many artists and critics."[18]

The fountain was designed as an allegory of Light and Water, combining the modern technologies of cast iron and gas lighting with the classically inspired fountain. The crenellated mural crown at the pinnacle is symbolic of a

---

[15] Jarrett, William Robertson Smith, p. 5.

[16] DeB. Randolph Keim, Keim's Illustrated Hand-Book of Washington and Its Environs: A Descriptive and Historical Hand-Book to the Capital of the United States of America, Washington: DeB. Randolph Keim, 1875, pp. 41–42.

[17] 19 Stat. 356.

[18] "Also on exhibit at the 1876 Exhibition was a part of one of Bartholdi's most famous works—the sixteen-foot-high right hand of the Statue of Liberty." Robert C. Byrd, United States Senate, "The Botanic Garden and Capital Landscape," Daily Edition, Congressional Record, v. 127, January 29, 1981, p. S803.

The Bartholdi Fountain in its present site in the Botanic Garden Park before restoration.

city wall. Bartholdi was unsuccessful in selling copies of the fountain to cities across the nation.

The fountain was moved from its location directly in front of the Capitol in 1927, when it was carefully dismantled; five years later, it was installed in its present location near the new Botanic Garden. The gas lamps were first lighted electrically in 1881, making the fountain a major attraction. Round glass globes were added when the fountain was completely electrified in 1915. The cast iron of the fountain, originally painted to look like bronze, has been regularly protected with coats of paint.

In 1986, the fountain was restored by the Architect of the Capitol in consultation with experts. After having been sandblasted, the fountain was given three layers of special epoxy-based coatings to protect it for many years to come. The basin was leveled so that the water flowed evenly, and the fountain was given new plumbing and electricity. Eventually, the Architect of the Capitol hopes to replace the light fixtures with reproductions of the original gas lamps.[19]

Conservatories

The three-hundred-foot-long Main Conservatory, begun in 1867 from designs by Edward Clark, the Architect of the Capitol, consisted of a central dome and two wings. In addition, there were ten smaller conservatories of brick and wood, one of which was a lecture hall, or botanical classroom capable of accommodating up to one hundred students. There were also four conservatories for propagating plants for distribution to Members of Congress.

Botanical Collections

The Garden's botanical collections were arranged according to a geographical distribution. Plants occupied the center Conservatory. Semi-tropical plants requiring protection were placed in one wing if they were found north of the equator, while those indigenous to countries lying south of the equator were placed in the other wing.

Occupying the center building, or rotunda, were more than three hundred different kinds of majestic palms as well as a wide variety of plants from Ceylon, China, India, Japan, Madagascar, New Zealand, Panama, and South America. The east wing was devoted primarily to the plants of the islands of the South Seas, Australia, Brazil, Cape of Good Hope, and New Holland. Plants in the west wing were from China, Japan, and the East and West Indies. Some of these plants were collected by Commodore Matthew

---

[19] Barbara A. Wolanin, Report From the Capitol Curator, The Capitol Dome, v. 21, n. 2, May 1986, p. 8.

Perry's U.S. Exploring Expedition, which was conducted from 1852 to 1855.[20]

A NEW PLAN FOR THE CAPITAL CITY AND ITS BOTANIC GARDEN

By the turn of the century, the United States Botanic Garden had won national prominence, but it had not yet found a permanent home. On March 8, 1901, Senator James McMillan secured passage of a resolution directing the Committee on the District of Columbia to study and prepare a report on "plans for the development and improvement of the entire park system of the District."[21] This resolution was to have a far-reaching effect on the Garden, ultimately resulting in the relocation to its present site.

Eleven days later, the Senate District Committee selected architect Daniel H. Burnham, of Chicago, and landscape architect Frederick Law Olmsted, Jr., of Massachusetts to propose plans for the development and improvement of the District's park system. Burnham and Olmsted in turn invited New York architect Charles F. McKim and sculptor Augustus Saint-Gaudens to work with them in preparing their plans. During the course of their investigation, the four men undertook an extensive tour of the United States and Europe in search of ideas they might incorporate in their final presentation.[22]

Ten months later, on January 15, 1902, the recommendations of the "Park Commission of the United States," as the four men were formally designated, were announced by the Senate Committee on the District of Columbia.[23] That afternoon, several Members of Congress met with President Theodore Roosevelt and his Cabinet to examine the elaborate plans that were on display at Washington's Corcoran Gallery.

---

[20] Keim, Keim's Illustrated Hand-Book of Washington, pp. 42–44. For Keim's full description, see Appendix 6. A list of the plants believed to have been presented to the Botanic Garden from the Perry Expedition is included as Appendix 7.

[21] Park System in the District of Columbia, Remarks in the Senate, Congressional Record, v. 35, March 8, 1901, p. 30.

[22] Clarence O. Sherrill, The Grant Memorial in Washington, Washington: U.S. Govt. Print. Off., 1924, p. 30.

[23] U.S. Congress, Senate, Committee on the District of Columbia, The Improvement of the Park System of the District of Columbia, Senate Report No. 166, 57th Cong., 1st Sess., Washington: U.S. Govt. Print. Off., 1902, 171 pp. (Serial No. 4258).

Included were two models, one showing Washington as it was and one showing what it might become.²⁴

One of the most dazzling features of the numerous changes proposed by the Park Commission was the creation of an open Mall between the Capitol and the Washington Monument, flanked on both sides by permanent public buildings as it is today. The Commission's plan amounted to an endorsement of the plan that Pierre Charles L'Enfant had developed in the 1790s.²⁵ "Gone would be the tangles of shrubbery, the Baltimore and Potomac Railroad Station and sheds littering the area, and the infamous area of crime known as Murder Bay." ²⁶

The location of the Botanic Garden was a problem for two reasons. First, the Garden and its greenhouses obstructed the planned vista between the Capitol and the Washington Monument. Equally problematic was the fact that the plan called for the large memorial to President Ulysses S. Grant, which Congress had approved in January of 1902,²⁷ to be erected on the Garden's existing site. The planners claimed that the Grant Memorial was the keystone of their design for an open Mall.²⁸

A sizeable number of Washingtonians, as well as Members of Congress and other governmental officials, openly opposed the Park Commission's plan because the placement of the Grant Memorial meant uprooting the magnificent trees on the Botanic Garden property. This protest was spearheaded through public appeals from Superintendent William R. Smith along with editorials by the Washington *Evening Star* denouncing the proposed destruction of the living memorials.²⁹ "Among the en-

---

²⁴ Charles Moore, Washington Past and Present, New York: The Century Co., 1929, pp. 263–264.
²⁵ Improvement of the Park System of DC, Senate Report No. 166, pp. 23–26, 35–36.
²⁶ Byrd, Botanic Garden, p. S803.
²⁷ 32 Stat. 460.
²⁸ Improvement of the Park System of DC, Senate Report No. 166, pp. 41–42.
²⁹ Answers Last Call, Washington Sunday Star, July 7, 1912, p. 2, pt. 2; Approved by Public: Star's Protest Against Destruction of Park Trees, Washington Evening Star, October 6, 1907, p. 2; Grand Old Trees May Escape Axe, Evening Star, October 7, 1907, p. 1; Most Noted Trees Succumb to the Axe, Evening Star, October 5, 1907, p. 1; and New Site Proposed For Grant Statue, Evening Star, October 12, 1907, p. 2.

dangered trees was an oak grown from an acorn taken from the tree over Confucius' grave, memorial trees honoring Presidents Hayes and Garfield, and trees planted by, or honoring, several members of Congress."[30]

Initially, the protesters prevailed and a temporary restraining order was issued in October 1907 postponing the construction of the Grant Memorial indefinitely.[31] The controversy was to be drawn out for more than another decade, but in the end the Memorial was erected exactly as planned with a lone concession being made to the protesters—the relocation, in April 1908, of three memorial trees: the Crittenden oak, Shepherd elm, and Beck elm.[32]

Altogether, twenty-one years would elapse between congressional authorization of the Grant Memorial and the actual completion of the project. All the long delay did was postpone the inevitable relocation of the Botanic Garden. Early in the 1920s, more than 200 trees on the grounds of the Garden were destroyed, William Smith's little red-brick cottage was razed, and the greenhouses were dismantled to make way for the Grant Memorial.[33] Then in 1926, Congress provided for the removal of the remainder of the Garden's buildings and the Bartholdi Fountain.[34]

Five years later, in November 1931, the cornerstone was laid for the Garden's new conservatory at Maryland Avenue and First Street, Southwest. The following year the Bartholdi Fountain was taken out of storage and placed in the square across Independence Avenue from the Conservatory in what is today called Bartholdi Park.[35]

---

[30] Byrd, Botanic Garden, p. S803.
[31] Sherrill, The Grant Memorial, p. 52.
[32] Ibid., p. 53. Each tree is discussed in Appendix 4.
[33] Byrd, Botanic Garden, p. S803.
[34] 44 Stat. 932. See also U.S. Congress, Joint Committee on the Library, Enlarging and Relocating the United States Botanic Garden, Report to Accompany S. 4153, Senate Report No. 748, 69th Cong., 1st Sess., Washington: U.S. Govt. Print. Off., 1926 (Serial No. 8526); and U.S. Congress, Joint Committee on the Library, Enlarging and Relocating the United States Botanic Garden, Senate Document No. 208, 69th Cong., 2d Sess., Washington: U.S. Govt. Print. Off., 1927 (Serial No. 8713).
[35] The garden was named the Frédéric Auguste Bartholdi Park by the Joint Committee on the Library in 1985.

END OF AN ERA

During the interim, William R. Smith, who had grown increasingly weary of the fight to preserve the Garden at its original site, died on July 12, 1912.[36] He was succeeded briefly by Charles Leslie Reynolds, who was appointed Superintendent on July 15, 1912.[37] Reynolds had been with the Garden for 40 years, 30 of which he had spent as Assistant Superintendent.

A little more than a year later, Reynolds suffered a fatal heart attack while "trying to catch some mischievous boys who had thrown stones through one of the greenhouses" at the Garden.[38]

THE GEORGE WESLEY HESS YEARS (1913–1934)

That December, George W. Hess was appointed Superintendent after receiving "a rating of 100 per cent as a specialist in the growing of foreign plants and 98 per cent in general gardening" on a civil service examination.[39] Hess brought to the position a diversified background that provided him with workable knowledge of several different botanical disciplines. At an early age, Hess began working with John Saul, a leading Washington florist and nurseryman.

---

[36] Answers Last Call, Washington Sunday Star, July 7, 1912, p. 2, pt. 1. A year-and-a-half before his death, Smith was interviewed by Catherine F. Cavanagh for New Age Magazine. Afterwards, Cavanagh wrote of how the battle to stop the uprooting of the Garden's famous trees had actually had a detrimental affect on Smith's health. Cavanagh, A Great Champion of Burns and Masonry, pp. 62–63.

[37] George P. Wetmore, Chairman of the Joint Committee on the Library to Elliott Woods, Superintendent, U.S. Capitol Building and Grounds, July 15, 1912, Records of the Architect of the Capitol (hereafter cited as AOC).

[38] Pursues Bad Boys: Is Picked Up Dead, Washington Evening Star, August 14, 1913, p. 12. See also Dies in Chase of Boys, Washington Post, August 14, 1913, p. 1.

[39] Famous National Botanic Garden About to be Given a New Lease on Life, Washington Evening Star, October 11, 1914, p. 3, pt. 4. See also Luke Lea, Chairman of the Joint Committee on the Library to Elliott Woods, Superintendent, U.S. Capitol Building and Grounds, December 13, 1913, AOC. A Copy of the December 22, 1913, oath taken by Hess is found in Ibid.

Eight years after he became Superintendent of the Garden, George Hess' title was changed to that of Director.[40] During his twenty-eight years of service, Hess brought several new dimensions to the Garden. He placed particular emphasis on education, exhibitions, and plant distribution as he sought to provide the Nation with a garden that would stand unrivaled in this country, if not in the world.

Eventually, he left Washington and gained additional experience through several different horticultural and landscaping pursuits. For several years he served as the motivating force for the Boston Public Garden and later as superintendent of the cemetery at Waltham, Massachusetts. Hess also worked in Florida for the Department of Agriculture.

Hess took great delight in showing groups of school children through the Garden and arranged special exhibits for their benefit that he felt would be of particular interest. These included displays of plants used for medicinal purposes, plants mentioned in the Bible, and plants whose products were used in the home. Once these were prepared, he then invited teachers throughout the Washington area to bring their classes to the Garden to take advantage of these special offerings. Hess personally conducted the children through the Garden until interest in his programs grew to the point that it was physically impossible for him to do it by himself.

Numerous articles appeared in the *Washington Star* between 1910 and 1925 describing the special events occurring at the Garden and the many plants that might be found on permanent exhibition there. These stories gave Hess ample opportunity to share proudly the Garden's most unusual holdings with the *Star's* readers.[41] Given such favorable publicity, it is not surprising that attendance during the Hess years was consistently high, especially for the Sunday openings that he initiated in 1915.

Hess was also the first Director to hold regular seasonal displays. His initial poinsettia show in 1916 was a huge success, attracting 3,000 visitors on opening day. That success started a tradition of Yuletide displays that has continued to the present.

---

[40] 41 Stat. 431.

[41] Mary Hughes, U.S. Botanic Garden Shines Brightly in Shadows of Bureaucratic Jungle, Florists' Review, v. 168, September 28, 1978, p. 102.

During his more than two decades as head of the Garden, Hess traveled extensively throughout the Nation and frequently returned with new plants to add to the Garden's collections. In addition, he added new material through trades with other public gardens. Unfortunately, accession records do not appear to have been maintained, and the details of when different species were added to the Garden are not known. Only through various newspaper accounts of the period is it possible to glean some idea of the unusual and "freak plants" that were grown at the Garden during the Hess years.[42]

While Hess was Director, the Botanic Garden was also active in nationwide plant distribution through a system of congressional allotments. Each year, Senators and Representatives were given a box containing approximately 80 varieties of shrubs, trees, and plants from the Botanic Garden. Plants selected for inclusion in each box depended upon the climate of the member's home State. Hess enthusiastically supported this program, maintaining that whenever a constituent received these plants and included them in his garden, neighbors would be motivated to improve their properties as well, and the results would be a more beautiful America.

Preparation for the annual distribution was a year-long effort. The magnitude of the task facing the Garden's staff, Washington's *Evening Star* reported in January 1916, was a tremendous undertaking in view of the vast contrasts in climate found throughout the country:

> All year a force of men has been working at the National Botanic Garden to prepare for the annual distribution of plants to Senators and Representatives, now being made. During ten months of the year these men are actually busy with the propagation and care of trees, shrubs and vines for the Members of Congress. The other two months are spent in preparation for greenhouse activity.
>
> The fruit of their labors is spread broadcast over the country; and the United States has within its confines practically every variety of temperature, every physical characteristic known in the world,

---

[42] The plants named in various newspaper articles are compiled in Appendix 8.

which means that the flora of the country includes the most diversified types and that the Botanic Garden must propagate them all in order to cater to the different climates of the widely separated states.[43]

This, the *Star* pointed out, was "necessary if the congressmen [were] to be truly benefited by a plant distribution. . . . Obviously, it would be [a] ridiculous and impractical procedure if plants were sent out promiscuously." To avoid such mistakes, Superintendent George Hess was "confronted with a responsibility that [implied] a special understanding of gardening and a national viewpoint on the situation." [44]

Each spring, during the Hess years, hundreds of plants were set out along the walk leading from the Garden's Third Street entrance to the Grant Memorial on the axis of the Washington Monument and the Capitol. Numerous other tropical plants were moved from the conservatories onto the grounds.

Tourists entering the north gate of the Garden were "confronted by two monster crotons as sentinels, startling in the brightness of their variegated foliage." On the northwest lawn was "a huge bed of exotics, such as alligator pears, guavas, sapadilla and paw paws." A plot of towering banana trees was found on the northeast lawn. "The most conspicuous adornment of the southwest lawn of the garden [was] a monster bed of various grasses, with the giant bamboo in the center surrounded by vari-colored and spiked and tufted grasses." Other interesting grasses included the "Bulrush of the Nile [*Cyperus papyrus*], in which Moses is recorded to have been hidden; the Andropogen [*Cymbopogon nardus*], from which is made citronella, used to drive away mosquitoes, and the attractive zebra grasses [*Miscanthus sinensis 'Zebrinus'*]." Along the "walk leading to the south gate [were] two glorious beds of scarlet and pink geraniums, which many floriculture experts [had] declared to have the most beautiful blooms in Washington."[45]

---

[43] Annual Distribution of Plants From the Botanic Garden, Washington Evening Star, January 30, 1916, p. 3, pt. 4. A list of the plants distributed to Congress in 1930 is found in Appendix 9.

[44] Ibid.

[45] G.W. Hess Returns From Southland, Washington Evening Star, August 19, 1915, p. 10. A description of what the Garden was like in the early 1920s is found in R.W. Shufeldt, Trees and Flowers in the United States Botanic Garden, American Forestry, v. 28, April 1922, pp. 226–231.

Each autumn, the palms and other tropical plants had to be moved back inside the conservatories. As the Garden's collections continued to grow, the problem of providing adequate space became a serious concern. Although the planning for the Grant Memorial was the primary motivation for proceeding with the plans for relocation of the Garden, the realities of limited space proved to be an equally persuasive argument.

DEBATE OVER RELOCATION

In 1916, Representative James L. Slayden, chairman of the Joint Committee on the Library, proposed that the Garden be relocated to Rock Creek Park and control of the facility be transferred to the Department of Agriculture. He felt that in a restricted urban setting the Garden could not possibly develop into a practical national organization. A spacious site such as Rock Creek Park, however, would allow for the development of an arboretum and the evolution of an institution befitting the name United States Botanic Garden.[46]

Those opposed to Representative Slayden's bill argued that the expense of a large site in Rock Creek Park would be too great and that it would be far less accessible to the public. Instead, they argued that the Garden should stay exactly where it was even though such a proposal would thwart the plans of the Park Commission. Conceding that space was a problem, they proposed adding 18.3 acres to the west.[47]

As it turned out, neither side was to have its way. An acceptable alternative, however, did ultimately result. In 1920, the Joint Committee on the Library, through a hearing, was able to make a careful examination of the condition of the

---

[46] See also U.S. Congress, House, Joint Committee on the Library, Botanic Garden, Report to Accompany H.R. 15313, House Report No. 641, 64th Cong., 1st Sess., Washington: U.S. Govt. Print. Off., 1916, 4 pp. (Serial No. 6904); and U.S. Congress, House, Joint Committee on the Library, Removal of the Botanic Garden: Minority Views, Report to Accompany H.R. 15313, House Report No. 642, Part 2, 64th Cong., 1st Sess., Washington: U.S. Govt. Print. Off., 1916, 3 pp. (Serial No. 6904).

[47] See also U.S. Congress, Joint Committee on the Library, Botanic Garden, Report to Accompany S. 6227, Senate Report No. 671, 64th Cong., 1st Sess., Washington: U.S. Govt. Print. Off., 1916, 2 pp. (Serial No. 6899).

Garden and gather valuable information regarding other botanic gardens throughout the world.[48]

Five years later, in January 1925, legislation was approved granting the Joint Committee on the Library authority "to investigate and report to Congress, with estimate of cost as to a new location for the conservatories of the United States Botanic Garden, south of the Mall in the vicinity of the present location." To complete its report, the Committee was "authorized to procure advice and assistance from any existing government agency, including the services of engineers, surveyors, draftsmen, landscape architects, and other technical personnel" in the various executive departments and independent agencies.[49]

Under the provisions of this Act, careful consideration was given to the property that would be needed, the most appropriate location, and what types of buildings would be most suitable.[50] The Committee's exploratory efforts culminated with the passage in January 1927 of legislation providing for the enlargement and relocation of the Garden and the Bartholdi Fountain to the sites they occupy today.[51]

Two months later, the Architect of the Capitol was authorized, under the direction and supervision of the Joint Committee on the Library, to provide for the construction of the new conservatories and other buildings that were to be constructed.[52] Between 1928 and 1932 the Architect of the Capitol expended $981,140.37 to clear the site of the old Botanic Garden, acquire a new parcel of land, and erect the new facilities.[53]

The site finally selected for the new Botanic Garden Conservatory was directly opposite the former site on the south side of Maryland Avenue. This new location fit very well into the Park Commission's plan, which had proposed that the Mall be flanked by public buildings.

---

[48] U.S. Congress, Architect of the Capitol, Annual Report of the Architect of the Capitol, 1927, Washington: U.S. Govt. Print. Off., 1928, p. 20; and U.S. Congress, Joint Committee on the Library, Establishment of a National Botanic Garden, Part 2, Hearings on S. 497 and S. Res. 165, 66th Cong., 2d Sess., Washington: U.S. Govt. Print. Off., 1920.

[49] 43 Stat. 729.

[50] Annual Report of the Architect of the Capitol, 1927, p. 20.

[51] 44 Stat. 931.

[52] 44 Stat. 1262.

[53] Architect of the Capitol, Annual Report of the Architect of the Capitol, 1937, Washington: U.S. Govt. Print. Off., 1937, p. 48. See also U.S. Botanic Garden: Preliminary Report, p. 17.

Senator Simeon D. Fess laying the cornerstone of the present Botanic Garden Conservatory on November 12, 1931. The four gentlemen accompanying him are, from left to right, Horace D. Rouzer, Assistant Architect of the Capitol; George W. Hess, Director of the Botanic Garden; Eugene Pugh of the George A. Fuller Co.; and David Lynn, Architect of the Capitol.

CHAPTER IV

THE PRESENT UNITED STATES BOTANIC GARDEN

A NEW LOCATION AND A NEW CONSERVATORY

On November 12, 1931, Senator Simeon D. Fess of Ohio, chairman of the Joint Committee on the Library, helped lay the cornerstone of the new Botanic Garden Conservatory at the foot of Capitol Hill.[1] The rectangular Conservatory, which has been in continual use since January 13, 1933, is approximately 183 feet by 262 feet and contains 47,674 square feet of floor space.

It was designed under the general supervision of David Lynn, Architect of the Capitol. The Chicago firm of Bennett, Parsons and Frost served as consulting architect. Structural engineer for the project was Louis E. Ritter of Chicago, and the New York firm of George A. Fuller Co. served as general contractor. Glazing for the Conservatory was done by the Lord & Burnham Co. of Irvington, New York. The total cost for constructing the Conservatory was $633,585.[2] A special feature of the conservatory is that it was the first large building to use aluminum for structure.[3]

The main or north entrance of the building, which is approximately 40 feet high, is made of limestone and has a flat concrete roof. At the top of the facade are four different keystones: Pan (a satyr head with horns, surrounded by wild flowers and oak leaves), Pomona (a female head with a band around it, surrounded by wild flowers), Triton (a male seagod head with aquatic flowers, water drops, and shells), and Flora (a young, smiling female face with roses). The front portion of the building, which is 17 feet by 200 feet, is

---

[1] According to an article in the Washington Star the following day, a box containing a record of the legislation pertaining to the Garden's improvement and a brief history of the Garden was placed inside the cornerstone, but a copy of these materials has not been found. Garden Cornerstone Laid, Washington Evening Star, November 13, 1931, p. B1.

[2] Annual Report of the Architect of the Capitol, 1937, p. 48.

[3] Federal Conservatory Uses Aluminum Alloy Framing, Engineering News-Record, April 14, 1932, pp. 539–542; and Hughes, U.S. Botanic Garden, p. 102.

The dome of the original Main Conservatory showing the cupola surrounded by a balustrade. Note the use of succulent members of the lily family in the bed in front of the building.

composed of an entrance hall and eleven sets of double doors large enough to allow the transfer of specimen plants to the patio in front during milder weather.

Inside the conservatory, there are eight garden rooms under glass, totaling 28,944 square feet of growing space. Seven smaller rooms surround the largest section, which occupies the central portion of the building known as the

Great Palm House. The domed roof of the Palm House is 27 feet high and 67 feet in diameter and is approximately 80 feet from the floor to the ceiling at the highest point. The Palm House is flanked on the east and west sides by open courtyards, each containing 4,992 square feet. The Conservatory also contains two 2,673-square-foot display wings, which are not under glass. These sections are used for flower shows, lectures, and special events.

Each section or room of the Conservatory is devoted to a display of different plant groups. Originally the eight rooms were named the Bay-Tree House, Succulent House, Succulent House Annex, Orangery, Orangery Annex, Fern House, Tropical Fruit House, and Palm House. As the Garden's collections have changed, so have the designations attached to these rooms. They presently are named Cycad House, Fern Connecting House, Fern House, Palm House, Cactus House, Cactus Connecting House, Bromeliad House, and Subtropical House. Inside each section of the building are planting boxes, which run the length of the room and vary in width from 4 to 15 feet.[4] The beds are separated by brick walkways varying in width from 4 to 6 feet. All of the beds are edged with calcareous tufa rock.[5]

At the same time the new Conservatory was being built, a Director's home was constructed at the rear of the Garden. This house replaced the two-story cottage on the old Botanic Garden grounds in which both William R. Smith and George W. Hess lived while in charge of the Garden. In 1933, the *Washington Star* depicted the Director's new home as being "very 'bijou,'" a "jewel indeed, unique, expensive, exclusive, magnificent and historic."[6] Today, the eight-room house serves as the Garden's office and is the work place of the Executive Director and the administrative staff.

---

[4] Details on the history of the Conservatory can be found in the records of the Architect of the Capitol.

[5] The calcareous rock used in the Botanic Garden came from Centralia, Ohio. It was formed in the bed of a highly charged subterranean lime spring, which emerged from the ground in that area. The water in the spring was dammed by beavers and an artificial lake was created. The lime precipitated to the bottom of the lake, along with many years' worth of decomposed organic matter. After the dam broke and this material was exposed to the elements it hardened. This material is known as tufa rock.

[6] Josephine Tighe Williams, This is the House that Uncle Sam Built, Sunday Star Magazine, May 7, 1933, pp. 3, 7.

Inside the Subtropical House of the Botanic Garden's Conservatory shortly after it was first planted. The drinking fountain in the foreground has since been removed.

Near the house, there was a smaller (4,000-square-foot) conservatory built in 1923 that was used for exhibiting citrus plants, cycads, crotons, camellias, and other cool house plants. This conservatory, designed and built by the American Greenhouse Manufacturers Company, contained a rockery with a pool, free planting areas, and a show house. Another two-story structure on the property was used by the Garden as a carpenter shop and storage shed. Both structures were removed in the early 1950s after having badly deteriorated.

DEVELOPMENT OF THE BOTANIC GARDEN PARK

It was while the Garden was being relocated early in the 1930s that the triangular one-acre plot across Independence Avenue just south of the new Conservatory was developed as an outdoor garden. Today, the Frédéric Auguste Bartholdi Park of the United States Botanic Garden Park is still used to display hardy plant material and seasonal floral exhibits. This Washington showplace features a wide variety of summer blooming annuals, rock garden perennials, and unusual trees and shrubs. The focal point of the park is the historic Bartholdi Fountain. This graceful thirty-foot-high cast-iron fountain, with its adorning aquatic monsters, fish, and caryatids, is a perfect ornament for the park.

NEW GROWING AREAS

Until the early 1960s, the Garden used nine greenhouses of various types and sizes as well as four cold frames adjacent to the Conservatory. These structures were utilized both for growing plant material to be displayed in the Conservatory and on the Garden's grounds and for maintaining a fairly extensive collection of orchids. The other buildings on the property, which the Garden began using for the first time in 1873, were storehouses, garages, and potting and working sheds. There was also a picturesque old two-story house with a mansard roof, built around 1842, which was used as a staff headquarters.

By 1956, however, the greenhouses next to the Conservatory had deteriorated so badly that they were no longer adequate for use and Congress authorized their demolition.[7]

---

[7] 70 Stat. 1068; and 72 Stat. 450. See also U.S. Senate, Committee on Rules and Administration, Replacement of Facilities of the Botanic Garden, Senate Report No. 2382, 84th Cong., 2d Sess., Washington: U.S. Govt. Print. Off., 1956, 2 pp. (Serial No. 11889); U.S. Congress, House, Committee on Public Works, Authorizing the Demolition and Removal of Certain Greenhouses and Other Structures . . . at the Botanic Garden Nursery, House Report No. 2878, 84th Cong., 2d Sess., Washington: U.S. Govt. Print. Off., 1958, pp. 1–2 (Serial No. 11901); Liz Hillenbrand, Old Botanic Garden to Vanish From Scene, Washington Post, August 13, 1956, p. 19; Demolition and Removal of the Greenhouses, Buildings, and Other Structures From Square 576 West Located Adjacent to the Site of the Main Conservatory and Construction of New Greenhouses and Service Buildings at Poplar Point Nursery in Replacement of Such Structures, February 16, 1956, p. 4. Botanic Garden Files, AOC.

Five years later, the plants that had been housed there were transferred to new structures at Poplar Point.[8]

Amidst the uproar over the relocation of the Garden that preceded the construction of the new Conservatory, 14.75 acres of land at Poplar Point in Anacostia, less than two miles from the Conservatory, was acquired in June 1926 as an additional growing area. Subsequently, eight greenhouses and a boiler room were constructed on this property. The Poplar Point site was enlarged by 9.87 acres in 1935 through the transfer of land originally held by the National Park Service.[9] This land has been used for the propagation and growing of plant material needed in and around the Conservatory as well as in landscaping the Government buildings on Capitol Hill.

Presently, there are 24 greenhouses at Poplar Point, comprising a total growing area of more than 50,000 square feet under glass. All of the plants used at the Garden's annual shows, the bedding materials used in the Garden's Park and on the grounds around the Conservatory, and the plant material used for landscaping the buildings on Capitol Hill are raised at the Garden's Anacostia site. The site also includes an outdoor nursery in which woody plants are raised for the grounds around the Capitol, congressional office buildings, Supreme Court, and Library of Congress.

On July 3, 1984, Congress enacted Public Law 98–340, 89 Stat. 308 ("Act") to facilitate the extension of the Washington Metropolitan Area Transit Authority "Green Line" through Anacostia. The extension project will entail the construction of a station, parking garage, surface parking areas, access roads, and related Metro facilities on land currently occupied by the United States Botanic Garden Nursery at Poplar Point. The Act transferred jurisdiction over the Poplar

---

[8] Karen Duffy, Removal and Relocation of Botanic Garden Greenhouses, February 11, 1971, Greenhouse File, Architect's Manuscript Files. See also Congress too Busy to Save Old House, Washington Evening Star, October 16, 1962, p. A6; and Mid-19th Century House With Mansard Roof in Sq. 576 West Removed, August 16, 1962, Mansard Roof House File, Architect's Manuscript Files.

[9] 44 Stat. 774; 47 Stat. 161; and Annual Report of the Architect of the Capitol, 1937, p. 48. See also PPN [Poplar Point Nursery] File, Architect's Manuscript Files.

Point Nursery to the Secretary of Interior and required the District of Columbia Government to take necessary action to convey to the Architect of the Capitol (without payment or consideration) certain real property located at "D. C. Village" to be used for the relocation of the Poplar Point Nursery. The District of Columbia is also required to effect the relocation of the Poplar Point Nursery and all of its facilities to the D. C. Village site under the provisions of the Uniform Relocation Assistance and Real Property Acquisition Policies Act of 1970, P.L. 91–646, 84 Stat. 1894, 42 U.S.C. 4601 (1976). Pending relocation, the Architect retains the right to continue the current use of the Poplar Point Nursery until he has approved, accepted, and occupied the relocated facility at D.C. Village. The new D.C. Village Facility will have 89,000 square feet of glasshouses, 30,000 square feet of support buildings, 15,000 square feet of cold frames, 4,900 square feet of lath houses, and 3 acres of irrigated growing fields.

ADMINISTRATION OF THE GARDEN

Closely coinciding with the completion of the new Conservatory and the initial development of the Poplar Point facility was a significant change in the actual administration of the Garden. Three days after George W. Hess' retirement on June 30, 1934,[10] Architect of the Capitol David Lynn was appointed Acting Director by Senator Alben W. Barkley, chairman of the Joint Committee on the Library, subject to the control, direction, and supervision of the Committee.[11]

---

[10] A Brief History of the United States Botanic Garden, July 20, 1934, p. 4, Historical File, Botanic Garden Files, AOC. Proposed legislation to make George W. Hess Director Emeritus at a salary of "$3,000 per annum, payable monthly out of the funds appropriated for expenses of the Botanic Garden" was passed by the Senate in March 1934. George W. Hess, Remarks in the Senate, Congressional Record, v. 78, March 29, 1934, p. 5740. The House, however, after a somewhat heated debate, chose not to approve S. 1839. George W. Hess, Remarks in the House, Congressional Record, v. 78, May 21, 1934, pp. 9199–9203. See also U.S. Congress, House, Committee on the Library, To Retire George W. Hess, House Report No. 1286, 73d Cong., 2d Sess., Washington: U.S. Govt. Print. Off., 1934, 2 pp. (Serial No. 9781); and U.S. Congress, House, Committee on the Library, George W. Hess, House Report No. 1325, Washington: U.S. Govt. Print. Off., 1934, 1 p. (Serial No. 9781).

[11] Alben W. Barkley, chairman of the Joint Committee on the Library, to David Lynn, Architect of the Capitol, July 3, 1934. AOC, Manuscript Files. For background on Lynn's appointment see John G. Bradley to Frederick A. Delano, July 3, 1934. Botanic Garden, F.A. Delano File, Manuscript Files. See also Legislative Appropriations, Remarks in the Senate, Congressional Record, v. 81, August 15, 1937, pp. 3509–3510 for a discussion on how the decision to make the Architect of the Capitol the Director of the Garden was viewed as a money-saving measure.

Lynn continued in this capacity until his retirement twenty years later. At the time of his appointment, Lynn was already well acquainted with the workings of the Garden, since he had been responsible for expenditures of nearly $1 million in connection with the Garden's relocation.[12]

Lynn's successor, J. George Stewart, was Architect of the Capitol and Acting Director of the Botanic Garden from September 1954 to May 1970.[13] Following Stewart's death on May 24, 1970, Mario E. Campioli served as Acting Architect of the Capitol and Acting Director of the Garden until the current Architect of the Capitol, George M. White, was appointed by President Richard M. Nixon on January 27, 1971.[14] After the Architect of the Capitol was appointed Acting Director, the Botanic Garden's Assistant Director took over the function of running the day-to-day operations of the organization.

The first Assistant Director of the Garden, Wilmer J. Paget, held the position from 1914 until 1945 (serving first under George W. Hess before his retirement in 1934, and then under Architect of the Capitol David Lynn). Paget is remembered by those who worked for him as an unusually gifted and knowledgeable plantsman, particularly in the field of plant taxonomy. He was well versed in a variety of horticultural fields and is remembered as the "brains" behind the cultivation and display of plants during his tenure. Paget first worked under his father, John Paget, a landscape gardener, at the Pennsylvania State Hospital before moving to Washington in 1901 to join the Botanic Garden staff as a laborer.[15]

---

[12] A Brief History of the United States Botanic Garden, July 20, 1934, pp. 4–5; and History of the United States Botanic Garden, May 31, 1938, p. 13. Botanic Garden Files, AOC.

[13] Frank A. Barrett, chairman of the Joint Committee on the Library, to J. George Stewart, September 25, 1954. AOC.

[14] White actually assumed responsibility for the two positions on February 11, 1971. U.S. President, 1968–1974 (Nixon), Architect of the Capitol, Announcement of Appointment of George Malcolm White of Ohio, January 29, 1971. Weekly Compilation of Presidential Documents, v. 7, February 1, 1971, p. 123; Wayne L. Hayes, Chairman of the Joint Committee on the Library to George M. White, Architect of the Capitol, February 11, May 3, June 10, July 8, 1971.

[15] Wilmer Paget: Horticulturist at Botanic Garden 41 Years, Washington Evening Star, April 26, 1968, p. B4; and Wilmer Paget Dies: D.C. Horticulturist, Washington Post, April 26, 1960, p. B2.

Upon his retirement Paget was succeeded by Edmund Emil Herman Sauerbrey, who began his career with the Botanic Garden in 1940. He was promoted to Assistant Director five years later and held the position until his death in 1968. Sauerbrey, a native of Thüringen, Germany, studied horticulture and landscaping in several German cities before emigrating to the United States in 1908.

While serving as general foreman and propagator for the Towson Nurseries in Baltimore, he developed a new double pink flowering lilac named in his honor. In 1938, he became an assistant botanist with the Departments of Agriculture in North Carolina and Georgia. Sauerbrey is remembered by those who worked with him at the Garden as a keen botanist who was rarely stumped by questions, whether botanical or horticultural in nature.[16]

Jimmie L. Crowe became the Garden's Assistant Director in October 1968, after six years as the horticulturist for the Capitol Grounds.[17] During his seventeen-year tenure, Crowe started successful programs such as the summer terrace displays, self-guided tours, and horticulture classes for the public. In addition to his years of service on Capitol Hill, Crowe served with the U.S. Air Force from 1945 to 1953, participating in the Berlin Airlift and the Korean War. After returning to the United States he earned a degree in horticulture from North Carolina State University before moving to Washington in 1961.[18]

Among the numerous other men and women who have contributed to the continuing excellence of the United States Botanic Garden, Albert T. DePilla's service is particularly noteworthy. On October 22, 1983, DePilla celebrated the

---

[16] Edmund E. H. Sauerbrey, Library of Congress Information Bulletin, v. 27, October 17, 1968, p. 633; [Edmund Sauerbrey] Top Official at Botanic Garden Here, Washington Post, October 5, 1968, p. B6; and Sauerbrey Rites Set Tomorrow, Washington Evening Star, October 3, 1968, p. B5.

[17] Architect of the Capitol to Jimmie L. Crowe, October 14, 1968, AOC.

[18] Silvio O. Conte, The Death of Assistant Director of the Botanical Gardens, Jimmie L. Crowe, on June 9, 1984, Remarks in the House, Congressional Record, Daily Edition, v. 130, June 12, 1984, p. E2757. See also Carol H. Flak, Are Your Gardens Grumpy? Just Call the Botanic Garden, Wall Street Journal, January 2, 1973, pp. 1, 18.

completion of his sixtieth year of Federal service. He began as a laborer with the Garden in 1923, at the age of sixteen, and except for a three-year-and-five-month tour of duty with the Army during World War II, DePilla spent his entire adult life as a member of the Garden's staff.[19]

The remarks of Senator Charles McC. Mathias, Jr., Chairman of the Joint Committee on the Library, on the anniversary of DePilla's sixtieth year of Federal service preceded the Senate's adoption of S. Res. 246 congratulating the personable botanist for his enduring contribution.[20]

With the appointment of David T. Scheid as Executive Director in July 1985, the United States Botanic Garden entered a time of change. Scheid's training included a Master's degree from Longwood Garden's program in Arboretum and Botanic Garden Management; six years as Superintendent of Nemours, a restored estate in Wilmington, Delaware; and three years as Vice President for Horticulture at the New York Botanical Garden. He reorganized the structure of the Garden to include the Conservatory Division, Production Division, Maintenance Division, Public Programs Division, and Administrative Office. One of his major goals was to create a garden of artistically displayed botanical collections.

CURRENT FUNCTIONS OF THE GARDEN

Despite its various moves, the Garden has deviated little during the past century and a half from the purposes set forth by the congressional charter given to the Columbian Institute.[21] Today, in much the same language as in the original charter, the Garden is authorized to "collect, cultivate, and grow the various vegetable productions of this and other countries for exhibition and display to the public and for study material for students, scientists,

---

[19] Elizabeth Mooney, He Performs Visual Magic With Plants, Washington Star-News, November 3, 1974, p. D10; Diane McCormick, Keeper of the Nation's Flowers, Roll Call, June 16, 1983, p. 18; and E. de la Garza, Botanic Garden, Congressional Record, v. 119, November 13, 1973, pp. E36918–E36919.

[20] Charles McC. Mathias, Jr, Congratulating Albert T. DePilla, Congressional Record, v. 119, October 21, 1983, pp. S14495–S14496.

[21] Rathbun, The Columbian Institute, p. 11.

and garden clubs."[22] Such a continuity of effort is rare among independent government agencies.

During recent years, collection and display have come to mean much more than just setting up something for tourists to look at as they wander through the Garden's Conservatory. A special emphasis has been placed on developing educational programs that will allow the visitors to the Garden to become active participants.

Since 1976, the Garden has sponsored a series of free horticulture classes from September through May. Topics covered include orchids as houseplants, environmental gardening, flower arranging, holiday plants, growing vegetables in containers, growing bromeliads, annuals for the home garden, and other subjects in which the Garden has developed expertise. Those attending these courses receive a series of plant culture sheets and special publications prepared by the staff. The Garden also sponsors a nationwide plant information service that handles hundreds of inquiries each week on a wide variety of botanical and horticultural topics.[23]

The Garden's library now contains more than 1,200 volumes, which are available to members of the staff and to the public by appointment. The collections housed in the Garden's Conservatory and displayed in its Park offer students and botanists, as well as professional horticulturists, an opportunity to view first hand a broad range of rare and interesting specimens from all over the world.

Since the relocation of the Garden to its present site in 1933, there has been a continual effort to acquire new species for both its indoor and outdoor collections.[24] Purchases of

---

[22] U.S. General Services Administration, National Archives and Records Service, Office of the Federal Register, The United States Government Organization Manual 1984/85, Washington: U.S. Govt. Print. Off., 1984, p. 39.

[23] Flak, Are Your Gardens Grumpy, pp. 1, 18; and Hughes, U.S. Botanic Garden, pp. 141–142.

[24] The Annual Reports of the Architect of the Capitol for 1933–1947 include the names of the plants donated to the Garden by Government institutions such as the Department of Agriculture as well as by private collectors. Several of the reports contain extensive and diversified lists of newly acquired plant material, including orchids, succulents, and hardy plants. The reports also mention the plants actually purchased by the Garden.

Orchid display in the U.S. Botanic Garden Conservatory.

new material are made through commercial companies as well as private collectors. Other acquisitions are regularly made through trades with other public institutions, both in this country and abroad, and through a sizeable number of donations—especially orchids. The Garden does not trade specimens with private collectors, nor does it sell any items.[25]

Although the Garden has no membership and does not sell plants or seeds, it does share information and exchange

---

[25] Hughes, U.S. Botanic Garden, p. 142.

plants on an informal basis with other botanic gardens, arboreta, and government agencies such as the Smithsonian Institution.[26]

Four annual plant and flower shows are sponsored by the Garden—an activity dating from George W. Hess' tenure as Director. The first event of the year is the Spring Flower Show, which is held from Palm Sunday through Easter Sunday. The Summer Terrace Display is held on the patio in front of the Conservatory from mid-May through October. Hundreds of flowering and foliage plants in hanging baskets and containers highlight this event, which is the Garden's longest running and most widely acclaimed show. It is also the latest program addition, having begun in 1973.[27]

Late in October, the Garden's third event, the Chrysanthemum Show, opens featuring more than 2,500 plants of over 150 varieties. From mid-December through New Year's Day, the public is invited to the Garden's Annual Poinsettia Show.[28] The Garden's staff begins preparations for this spectacular display in May when nursery staff start cuttings from older plants. In addition, the Garden also hosts annual plant and flower shows sponsored by area garden clubs and plant societies.

Although the Garden has not dispersed plant material through congressional allotments to the home districts of Members of Congress for more than fifty years,[29] it does raise, at its Poplar Point Production Facility and Nursery, indoor plants that are lent to congressional offices on a limited and restricted basis.

To carry out its various activities, the Garden currently employs fifty-seven people, including an Executive Direc-

---

[26] Specific legislation covering the exchange of plants is found in 49 Stat. 471.

[27] An extensive file on the Botanic Garden's various flower shows is found in Botanic Garden Files, AOC.

[28] The poinsettia was named after Joel Poinsett, who served in Congress as a Representative from South Carolina from 1821 to 1825. Then between 1825 and 1829, he was President John Quincy Adams' Minister to Mexico, where the flaming red flower grew wild. When he returned to Washington he brought cuttings of the plant home with him. Byrd, Botanic Garden, p. S804. Poinsett was a member of the Columbian Institute.

[29] Congress ended this practice in 1934. 48 Stat. 828.

tor, two facility managers, a public programs specialist, a botanist, a small clerical staff, numerous gardeners, a maintenance crew, and several night watchmen.[30]

FUTURE DIRECTIONS OF THE GARDEN

An assessment of the physical condition of the facility is now under way as the first step in the restoration and retrofitting of the Conservatory. This would adapt it for another fifty years of service by strengthening and modernizing the physical structure and support systems. A new Master Plan, which brings appropriate treatment and improvements to each glasshouse, is gradually being implemented. In addition, using the original drawings, some of the display areas have been renovated to restore original features, such as the reflecting pool in the Palm House.

The Production Facility currently (1991) housed at Poplar Point, in Anacostia, Southeast, Washington, D.C., will be relocated to D. C. Village in Southeast, Washington, D.C., in 1993. The new facility, which will be larger than the old, will use the lastest greenhouse technology and will allow for greater refinement of the original intentions, serving the public and Congress.

Modern technology is now being introduced throughout the garden with the adoption of new growing techniques and the installation of computers at all locations. Themes have been added to all shows, and a 12-month horticultural exhibit, displaying botanically interesting and colorful plants, has been added to the Subtropical House.

Public outreach and education are making great strides as the topics and teachers for horticultural classes are upgraded to utilize prominent and knowledgeable individuals speaking on timely issues.

Congress has authorized the establishment of the National Garden at the United States Botanic Garden, commemorating the Bicentennial of Congress. The National Garden will display the many varieties of our national

---

[30] U.S. Congress, Senate, Committee on Appropriations, Legislative Branch Appropriations, 1985, Senate Report No. 98-515, 98th Cong., 2d Sess., Washington: U.S. Govt. Print. Off., 1984, p. 24.

flower, the rose, as well as many interesting and important plants, both native and introduced. This garden will be used for teaching and will be designed to be accessible to all who come to visit, work, and learn. For the first time, and in conjunction with the National Garden, Congress has authorized the acceptance of gifts to the Botanic Garden, including volunteer service. In January 1991 a private non-profit organization known as the National Fund for the United States Botanic Garden was chartered under District of Columbia law to raise funds for the National Garden. This group represents a modern version of the private efforts that led to the initial creation of the Botanic Garden. The two-and-one-half-acre site for the National Garden, located on the west end of the Conservatory, is a prime location for this new American educational display garden. It is hoped that this garden will inspire and educate the millions of visitors who will come to experience it.

In these ways, the United States Botanic Garden is positioning itself to fulfill its charter in response to the demands and challenges of the twenty-first century. At the same time, these efforts move the Garden closer to realizing the vision of the original Columbian Institute for the Promotion of Arts and Sciences.

CONCLUSION

The United States has changed dramatically since 1816, when the Columbian Institute for the Promotion of Arts and Sciences formed a botanic garden at the foot of Capitol Hill to collect, cultivate, and distribute useful plants to the American people. Some of the goals of the Columbian Institute are now shared by the United States Department of Agriculture and the Smithsonian Institution on a larger scale than was ever envisioned by the Garden's founding fathers.

The United States Botanic Garden continues to fulfill several important roles through its collection and cultivation of the world's flora. The throngs of visitors who pass through the Garden each year attest to the value and popularity of those functions. In 1978, it was recognized as the Nation's best maintained botanic garden or arboretum by the Professional Grounds Management Society.[31]

---

[31] Botanic Garden Best Maintained, Roll Call, March 16, 1978, p. 5.

The United States Botanic Garden, after almost 175 years of existence, is an institution that contains not only plants of historical importance to the nation, but collections vital to research in areas as diverse as medicine, the reforestation of tropical areas, and the preservation of species.

Through educational programs, exhibits, and plant information services, all citizens and visitors can make use of the Garden's resources, whether or not they live in the Nation's Capital.

With the continued support of Congress, the United States Botanic Garden will become the institution envisioned by George Washington—a garden that reflects the diversity of plants and their importance in human life.

CHAPTER V

A SELECTED BIBLIOGRAPHY

Ancestral Vine of all American Amelopsis in the Botanic Garden. Washington Sunday Star, Oct. 20, 1907. p. 6, pt. 4.

Annual Distribution of Plants from the Botanic Garden. Washington Evening Star, Jan. 30, 1916. p. 3, pt. 4.

Answers Last Call: Passing of William R. Smith at the Age of Eighty-Four. Washington Sunday Star, July 7, 1912. p. 2, pt. 1.

Approved by Public: Star's Protest Against Destruction of Park Trees. Washington Evening Star, Oct. 6, 1907. p. 2.

Barnhart, John Hendley. Brackenridge and His Book on Ferns. Journal of the New York Botanical Garden, v. 20, June 1919. pp. 117–124.

Bartlett, Harley Harris. The Reports of the Wilkes Expedition and the Work of Specialists in Science. Proceedings of the American Philosophical Society, v. 82, June 29, 1940. pp. 601–705.

Botanic Garden Best Maintained. Roll Call, Mar. 16, 1978. p. 5.

Brackenridge, W.D. Gardner's Age, v. 26, Dec. 1884. pp. 375–376.

_____. Meehan's Monthly, v. 3, Mar. 3, 1893. p. 47.

Bryan, G. S. The Purpose, Equipment and Personnel of the Wilkes Expedition. Proceedings of the American Philosophical Society, v. 82, June 29, 1940. pp. 551–560.

_____. The Wilkes Exploring Expedition. United States Naval Institute Proceedings, v. 65, Oct. 1939. pp. 1452–1464.

Bryan, Wilhelmus Bogart. A History of the National Capitol. 2 vols. New York: Macmillan Company, 1916.

Byrd, Robert C. The United States Senate [The Botanic Garden and Capitol landscape]. Remarks in the Senate, Congressional Record [daily ed.], v. 127, Jan. 29, 1981. pp. S801–S806.

Caemmerer, H. P. Washington: The National Capitol. Washington: U.S. Govt. Print. Off., 1932.

Cavanagh, Catherine Francis. A Great Champion of Burns and Masonry. New Age Magazine, v. 14, Jan. 1911. pp. 59–67.

Congress too Busy to save Old House. Washington Evening Star, Oct. 16, 1962. p. A6.

Conklin, Edward G. Connection of the American Philosophical Society with our first National Exploring Expedition. Proceedings of the American Philosophical Society, v. 82, June 29, 1940. pp. 519–541.

Cooley, Mark. The Exploring Expedition in the Pacific. Proceedings of the American Philosophical Society, v. 82, June 29, 1940. pp. 707–719.

Conte, Silvio O. The Death of Assistant Director of the Botanical Gardens, Jimmie L. Crowe, on June 9, 1984. Remarks in the House. Congressional Record [daily ed.], v. 130, June 12, 1984. p. E2757.

de la Garza, E. Botanic Garden. Remarks in the House. Congressional Record, v. 119, Nov. 13, 1973. pp. E36918–E36919.

Dies in Chase of Boys. Washington Post, Aug. 14, 1913. p. 1.

Editorial. Daily National Intelligencer, May 22, 1828. p. 3.

Edmund E. H. Sauerbrey. Library of Congress Information Bulletin, v. 27, Oct. 17, 1968. p. 633.

_____. Top Official at Botanic Garden Here. Washington Post, Oct. 5, 1968. p. B6.

Famous National Botanic Garden About To Be Given a New Lease on Life. Washington Evening Star, Oct. 11, 1914. p. 3, pt. 4.

Federal Conservatory uses Aluminum Alloy Framing. Engineering News-Record, Apr. 14, 1932. pp. 539–542.

Feipel, Louis N. The Wilkes Expedition: Its Progress through Half a Century: 1826–1876. United States Naval Institute Proceedings, v. 40, Sept.-Oct. 1914. pp. 1323–1350.

Flak, Carol H. Are your Gardens Grumpy? Just call the Botanic Garden. Wall Street Journal, Jan. 2, 1973. pp. 1, 18.

G. W. Hess Returns From Southland. Washington Star, Aug. 19, 1915. p. 10.

Garden Corner Stone Laid. Washington Star, Nov. 13, 1931. p. B1.

Goode, G. Brown. The Genesis of the National Museum. In Annual Report of the Board of Regents of the Smithsonian Institute for the year ending June 30, 1891. Washington: U.S. Govt. Print. Off., 1892. pp. 273–364.

Grand Old Trees May Escape Axe. Evening Star, Oct. 7, 1907. p. 1.

Gray, Jane Loring, ed. Letters of Asa Gray. New York: Lenox Hill Publishing and Distribution Co., 1893.

Green, Constance McLaughlin. Washington: Village and Capital, 1800–1878. 2 vols. Princeton, New Jersey: Princeton University Press, 1962.

Haskell, Daniel C. The United States Exploring Expedition 1838–1842. New York: New York Public Library, 1942.

Henderson, Daniel. The Hidden Coast: a Biography of Admiral Charles Wilkes. New York: William Sloan Associates Publishers, 1953.

Hess, George W. Remarks in the House. Congressional Record, v. 78, May 21, 1934. pp. 9199–9203.

_____. Remarks in the House. Congressional Record, v. 78, Mar. 29, 1934. p. 5740.

Hill, James D. Charles Wilkes—Turbulent Scholar of the Old Navy. United States Naval Institute Proceedings, v. 57, July 1931. pp. 867–887.

Hillenbrand, Liz. Old Botanic Garden to vanish from scene. Washington Post, Aug. 13, 1956. p. 19.

Hobbs, William Herbert. The Discovery of Wilkes Land, Antarctica. Proceedings of the American Philosophical Society, v. 82, June 29, 1940. pp. 561–582.

Hovey, C. M. Experimental Garden of the National Institute. Magazine of Horticulture, v. 10, Mar. 1844. pp. 81–83.

Hughes, Mary. U.S. Botanic Garden shines brightly in Shadows of Bureaucratic Jungle. Florists' Review, v. 163, Sept. 28, 1978. pp. 100–104, 141.

Introduction of Foreign Plants and Seeds. Daily National Intelligencer, Nov. 17, 1827. p. 2.

Jaffe, David. Literary Detective Harpoons a Whale of a Tale. Potomac Magazine (Washington Post), June 2, 1963. pp. 18–19.

_____. The Stormy Petrel and Whale: Some Origins of Moby Dick. Washington: University Press of America, 1982. pp. 7–38.

Jarrett, James MacPherson. William Robert Smith—Founder of the Saint Andrew's Society of Washington, D.C. Newsletter of the Saint Andrew's Society of Washington, D.C., No. 123, July 1974. pp. 4–5.

Johnson, Allen and Dumas Malone, eds. Dictionary of American Biography. 20 vols. New York: Charles Scribner's Sons, 1928–1937.

Keim, DeB. Randolph. Keim's Illustrated Hand-Book of Washington and its Environs: A Descriptive and Historical Hand-Book to the Capitol of the United States of America. Washington: DeB. Randolph Keim, 1877. pp. 41–42.

Legislative Appropriations. Remarks in the Senate. Congressional Record, v. 81, Aug. 15, 1937. pp. 3509–3510.

Letter from Columbia Institute. Daily National Intelligencer, Nov. 24, 1827. p. 2.

Mathias, Charles McC., Jr. Congratulating Albert T. DePilla. Remarks in the Senate. Congressional Record [daily ed.], v. 119, Oct. 21, 1983. pp. S14495–S14496.

MacHatton, Robert Park. Heritage of the Navy. United States Naval Institute Proceedings, v. 68, July 1942. pp. 967–969.

McCormick, Diane. Keeper of the Nation's Flowers. Roll Call, June 16, 1983. p. 18.

Mooney, Elizabeth. He Performs Visual Magic with Plants. Washington Star-News, Nov. 3, 1974. p. D10.

Moore, Charles. Washington Past and Present. New York: The Century Co., 1929. pp. 263–264.

Morgan, William James, David Tyler, Joye L. Leonhart, and Mary F. Loughlin, eds. Autobiography of Rear Admiral Charles Wilkes, U.S. Navy 1798–1877. Washington: Department of Navy, Naval History Division, 1978.

Morsberger, Robert E. The Wilkes' Expedition: 1838–1842. American History Illustrated, v. 7, June 1972. pp. 4–10, 45–49.

New Site Proposed For Grant Statue. Washington Evening Star, Oct. 12, 1907. p. 2.

Park System in the District of Columbia. Congressional Record, v. 35, Mar. 8, 1901. p. 30.

Pinkett, Harold T. Early Agricultural Societies in the District of Columbia. Records of the Columbia Historical Society, v. 51–52, 1951–1952. pp. 32–45.

Pursues Bad Boys: Is Picked Up Dead. Washington Evening Star, Aug. 14, 1913. p. 12.

Rathbun, Richard. The Columbian Institute for the Promotion of Arts and Sciences. Bulletin 101 of the Smithsonian Institution. Washington: U.S. Govt. Print. Off., 1917.

Shufeldt, R. W. Trees and Flowers in the United States Botanic Garden. American Forestry, v. 28, Apr. 1922. pp. 226–231.

Second Bulletin of the Proceedings of the National Institute for the Promotion of Science. Washington: printed by Peter Force, 1842.

Sherrill, Clarence O. The Grant Memorial in Washington. Washington: U.S. Govt. Print. Off., 1924. p. 30.

Smith, William R. A Catalog of Plants in the National Conservatories: A Popular Catalogue of the Extraordinary Curiosities in the National Institute Arranged in the Building Belonging to the Patent Office. Washington: Alfred Hunter, 1854.

Solit, Karen D. The U.S. Botanic Garden. American Horticulturist, v. 61, Apr. 1982. pp. 4–6.

Strauss, W. Patrick. Preparing the Wilkes Expedition: A Study in Disorganization. Pacific Historical Review, v. 28, Aug. 1959. pp. 221–232.

Sauerbrey Rites Set Tomorrow. Washington Evening Star, Oct. 3, 1968. p. B5.

U.S. Architect of the Capitol. Annual Report of the Architect of the Capitol, 1927. Washington: U.S. Govt. Print. Off., 1928.

_____. Annual Report of the Architect of the Capitol, 1937. Washington: U.S. Govt. Print. Off., 1937.

U.S. Architect of the Capitol. Office of the Curator. Manuscript Files.

U.S. Congress. House. Committee on the District of Columbia. Subcommittee on Fiscal Affairs. Relocation of the Architect's Tree Nursery, Hearings and Markups, on H.R. 4153 and H.R. 5565, 98th Cong., 1st and 2d Sess. Washington: U.S. Govt. Print. Off., 1984. 57 pp.

U.S. Congress. House. Committee on the Library. George W. Hess. House Report no. 1325. Washington: U.S. Govt. Print. Off., 1934. 1 p. (Serial no. 9781).

_____. United States Botanic Garden. Preliminary Report and Memorandum on the U.S. Botanic Garden and Kindred Institutions Together with Certain Recommendations looking through its Improvements as a Scientific, Educational, and Aesthetic Accomplishment. Committee Print, 73d Cong., 2d Sess. Washington: U.S. Govt. Print. Off., 1934. 22 pp.

_____. To Retire George W. Hess. House Report no. 1286, 73d Cong., 2d Sess. Washington: U.S. Govt. Print. Off., 1934. 2 pp. (Serial no. 9781).

U.S. Congress. House. Committee on Public Buildings. John McArann. House Report no. 290, 24th Cong., 2d Sess. Washington: Blair and Rives Printers, 1837. 5 pp. (Serial no. 306).

U.S. Congress. House. Committee on Public Buildings and Grounds. Conservatory and Other Buildings, and Additional Land, United States Botanic Garden. House Report no. 286, 68th Cong., 1st Sess. Washington: U.S. Govt. Print. Off., 1924. 2 pp. (Serial no. 8227).

U.S. Congress. House. Committee on Public Works. Authorizing the Demolition and Removal of Certain Greenhouses and Other Structures

... at the Botanic Garden Nursery. House Report no. 2878, 84th Cong., 2d Sess. Washington: U.S. Govt. Print Off., 1958. 2 pp. (Serial no. 11901).

U.S. Congress. Joint Committee on the Library. Botanic Garden. Report to accompany S. 6227. Senate. Report no. 671, 64th Cong., 1st Sess. Washington: U.S. Govt. Print. Off., 1916. 2 pp. (Serial no. 6899).

———. Botanic Garden. Report to accompany H.R. 15313. House. Report no. 641, 64th Cong., 1st Sess. Washington: U.S. Govt. Print. Off., 1916. 4 pp. (Serial no. 6904).

———. The Botanic Garden and its relation to the Joint Committee on the Library. July 1, 1912. Washington: U.S. Govt. Print. Off., 1912. 8 pp.

———. Enlarging and Relocating the United States Botanic Garden. Report to accompany S. 4153. Senate Report no. 748, 69th Cong., 1st Sess. Washington: U.S. Govt. Print. Off., 1926. 5 pp. (Serial no. 8526).

———. Enlarging and Relocating the United States Botanic Garden. Senate. Document no. 208, 69th Cong., 2d Sess. Washington: U.S. Govt. Print. Off., 1927. 2 pp. (Serial no. 8713).

———. Establishment of a National Botanic Garden. Part 2. Hearings on S. 497 and S. Res. 165, 66th Cong., 2d Sess. Washington: U.S. Govt. Print. Off., 1920. 152 pp.

———. Removal of the Botanic Garden: Minority Views. Report to accompany H.R. 15313. House Report no. 642. Part 2, 64th Cong., 1st Sess. Washington: U.S. Govt. Print. Off., 1916. 3 pp. (Serial no. 6904).

U.S. Congress. Senate. Committee on Appropriations. Legislative Branch Appropriations, 1985. Senate. Report no. 98-515, 98th Cong., 2d Sess. Washington: U.S. Govt. Print. Off., 1984. p. 24.

U.S. Senate. Committee on the District of Columbia. The Improvement of the Park System of the District of Columbia. Senate. Report no. 166, 57th Cong., 1st Sess. Washington: U.S. Govt. Print. Off., 1902. 171 pp. (Serial no. 4258).

U.S. Congress. Senate. Committee on Rules and Administration. Replacement of Facilities of the Botanic Garden. Senate Report no. 2382, 84th Cong., 2d Sess. Washington: U.S. Govt. Print. Off., 1956. 2 pp. (Serial no. 11889).

———. Estimates of expenditures on the Botanic Garden, 1850-1907. Senate. Document no. 494, 60th Cong., 1st Sess. Washington: U.S. Govt. Print. Off., 1908. 13 pp.

U.S. Library of Congress. Manuscript Division. Peter Force Papers, Series 8D.

———. Records of Proceedings of the Columbian Institute for the Promotion of Arts and Sciences (M358).

U.S. General Services Administration. National Archives and Records Service. Office of the Federal Register. The United States Government Organization Manual 1984/85. Washington: U.S. Govt. Print. Off., 1984. p. 39.

U.S. Statutes at Large. Vols. 1-98.

Viola, Herman J., and Carolyn Margolis, eds. Magnificent Voyagers: The U.S. Exploring Expedition, 1838-1842. Washington: Smithsonian Institution Press, 1985.

Wilkes, Charles. Narrative of the United States Exploring Expedition During the Years 1838, 1839, 1840, 1841, 1842. Philadelphia: Lea and Blanchard, 1845.

_____. Synopsis of the Cruise of the U.S. Exploring Expedition During the Years 1838, 1839, 1840, and 1842. Washington: Peter Force, 1842.

Williams, Josephine Tighe. This is the House That Uncle Sam Built. Sunday Star Magazine, May 7, 1933. pp. 3, 7.

Wilmer Paget Dies: D.C. Horticulturist. Washington Post, Apr. 26, 1960. p. B2.

Wilmer Paget: Horticulturist at Botanic Garden 41 Years. Washington Evening Star, Apr. 26, 1968. p. B4.

APPENDIX 1

CIRCULAR LETTER AND REPORT OF 1827

SECRETARY RICHARD RUSH'S LETTER (CIRCULAR) AND DR. J.M. STAUGHTON'S REPORT ON PLANT COLLECTION, PRESERVATION, AND SHIPMENT FOR THE COLUMBIAN INSTITUTE'S BOTANIC GARDEN.*

CIRCULAR

To a Portion of the Consuls of the United States

TREASURY DEPARTMENT

September 6, 1827

SIR:

The President is desirous of causing to be introduced into the United States all such trees and plants from other countries not heretofore known in the United States, as may give promise, under proper cultivation, of flourishing and becoming useful, as well as superior varieties of such as are already cultivated here. To this end I have his directions to address myself to you, invoking your aid to give effect to the plan that he has in view. Forest trees useful for timber; grain of any description; fruit trees; vegetables for the table; esculent roots; and, in short, plants of whatever nature whether useful as food for man or the domestic animals, or for purposes connected with manufactures or any of the useful arts, fall within the scope of the plan proposed. A specification of some of them to be had in the country where you reside, and believed to fall under one or other of the above heads, is given at the foot of this letter, as samples merely, it not being intended to exclude others of which you may yourself have knowledge, or be able, on inquiry, to obtain knowledge. With any that you may have it in your power to send, it will be desirable to send such notices of

---

* From the Daily National Intelligencer, November 17, 1827, p. 2.

their cultivation and natural history as may be attainable in the country to which they are indigenous; and the following questions are amongst those that will indicate the particulars concerning which information may be sought:

1. The latitude and soil in which the plant most flourishes.

2. What are the seasons of its bloom and maturity, and what the term of its duration?

3. In what manner is it propagated? by roots, seeds, buds, grafts, layers, or how? and how cultivated? and are there any unusual circumstances attending its cultivation?

4. Is it affected by frost, in countries where frost prevails?

5. The native or popular name of the plant, and (where known) its botanical name and character.

6. The elevation of the place of its growth above the level of the sea.

7. Is there in the agricultural literature of the country, any special treatise or dissertation upon its culture? If so, let it be stated.

8. Is there any insect particularly habituated to it?

9. Lastly—its use, whether for food, medicine, or the arts.

In removing seeds or plants from remote places across the ocean, or otherwise, great care is often necessary to be observed in the manner of putting them up and conveying them. To aid your efforts in this respect upon the present occasion, a paper of directions has been prepared, and is herewith transmitted.

The President will hope for your attention to the objects of this communication as far as circumstances will allow; and it is not doubted that your own public feelings will impart to your endeavours under it, a zeal proportioned to the beneficial results to which the communication looks. It is proper to add, that no expense can at present be authorized in relation to it. It is possible, however, that Congress may not be indisposed to provide a small fund for it. The seeds, plants, cuttings, or whatever other germinating substance you may transmit, must be addressed to the Treasury Department, and sent to the Collector of the port to which the vessel conveying them is destined, or where she may arrive, accompanied by a letter of advice to the department. The Secretary of the Navy has instructed the Commanders of such of the public vessels of the United States as may ever touch at your port, to lend you their assistance towards giving effect to the objects of this communication; as you will perceive by the copy of his letter of instructions, which is herewith enclosed for your information. It is believed, also, that the Masters of the merchant vessels of the United States, will generally be willing—such is their well-known public spirit—to lend their gratuitous co-operation towards effecting the objects proposed.

I remain, respectfully,

Your most obedient servant,

(Signed Richard Rush)

DIRECTIONS

For Putting Up and Transmitting

SEEDS AND PLANTS

(Accompanying the Letter of the Secretary of the Treasury of September 6, 1827)

With a view to the transmission of seeds from distant countries, the first object of care is to obtain seeds that are fully ripe, and in a sound and healthy state. To this the strictest attention should be paid; otherwise, all the care and trouble that may be bestowed on them, will have been wasted on objects utterly useless.

Those seeds that are not dry when gathered, should be rendered so by exposure to the air, in the shade.

When dry, the seeds should be put into paper bags. Common brown paper has been found to answer well for making such bags. But, as the mode of manufacturing that paper varies in different countries, the precaution should be used of putting a portion of the seeds in other kinds of paper. Those that most effectually exclude air and moisture, are believed to be the best for that purpose. It would be proper, also, to enclose some of the seeds in paper or cloth that has been steeped in melted bees-wax. It has been recommended that seeds collected in a moist country, or season, be packed in charcoal.

After being put up according to any of these modes, the seeds should be enclosed in a box; which should be covered with pitch to protect them from damp, insects, and mice. During the voyage they should be kept in a cool, airy, and dry situation;—not in the hold of the ship.

The oil seeds soonest lose their germinating faculty. They should be put in a box with sandy earth, in the following manner:—first about two inches of earth at the bottom, into this the seeds should be placed at distances proportionate to their size; on these another layer of earth about an inch thick; and then another layer of seeds;—and so on with alternate layers of earth and seeds until the box is filled within about a foot of the top, which space should be filled with sand; taking care that the earth and sand be well put in, that the seeds may not get out of place. The box should then be covered with a close network of cord well pitched, or with split hoops of laths also pitched; so as to admit the air without exposing the contents of the box to be disturbed by mice or accident. The seeds thus put up will germinate during their passage, and will be in a state to be planted immediately on their arrival.

Although some seeds with a hard shell, such as nuts, peaches, plums, &c, do not come up until a long time after they are sown, it would be proper, when the kernel is oily, to follow the method just pointed out, that they may not turn rancid on the passage. This precaution is also useful for the family of laurels, (*laurineae*) and that of myrtles (*myrti*) especially when they have to cross the equatorial seas.

To guard against the casualties to which seeds in a germinating state may be exposed during a long voyage, and, as another means of ensuring the success of seeds of the kinds here recommended to be put into boxes with earth, it would be well, also, to enclose some of them (each seed separately) in a coat of bees-wax, and afterwards pack them in a box covered with pitch.

In many cases it may be necessary to transmit roots. Where roots are to be transmitted, fibrous roots should be dealt with in the manner herein recommended for young plants. Bulbous and tuberous roots should be put into boxes in the same manner as has already been recommended for oleagenous seeds; except, that instead of earth, dry sand as free as possible from earthy particles, should be used. Some of the bulbous and tuberous roots, instead of being packed in sand, may be wrapped in paper, and put in boxes covered with net-work or laths. Roots should not be put in the same box with seeds.

Where the seeds of plants cannot be successfully transmitted, they may be sown in boxes, and sent in a vegetating state. Where more than one kind is sown in the same box, they should be kept distinct by laths, fastened in it crosswise on a level with the surface of the ground in which they are sown; and, when different soils are required, it will be necessary to make separate compartments in the box.

In either case they should be properly marked, and referred to in the descriptive notes which accompany them.

When plants cannot be propagated from seeds with a certainty of their possessing the same qualities which long culture or other causes may have given them, they may be sent in a growing state. For this purpose, they should be taken up when young. Those, however, who are acquainted with their cultivation in the countries where they grow, will know at what age they may be safely and advantageously removed. They may be transplanted direct into the boxes in which they are to be conveyed; or, where that cannot be conveniently done, they may be taken up with a ball of earth about the roots, and the roots of each surrounded with wet moss, carefully tied about it to keep the earth moist. They may afterwards be put into a box, and each plant secured by laths fastened crosswise above the roots, and the interstices between the roots, filled with wet moss. The same methods may be observed with young grafted or budded fruit trees.

Where the time will permit, it is desirable that the roots of the plants be well established in the boxes in which they are transplanted. Herbaceous plants require only a short time for this; but, for plants of a woody texture, two or three months is sometimes necessary.

Boxes for the conveyance of plants, or of seeds that are sown, may be made about two feet broad, two feet deep, and four feet long, with small holes in the bottom, covered with a shell, or piece of tile or other similar substance for letting off any superfluous water. There should be a layer of wet moss of two or three inches deep at the bottom, or, if that cannot be had, some very rotten wood or decayed leaves, and upon that about twelve inches depth of fresh loamy earth, into which the plants that are to be transplanted should be set. The surface of the earth should be covered with a thin layer of moss cut small, which should be occasionally washed in fresh water during the voyage, both to keep the surface moist, and to wash off mouldness or any saline particles that may be on it.

When the boxes are about to be put on board the ship, hoops of wood should be fastened to the sides, in such a manner, that, arching over the box, they may cover the highest of the plants; and over these should be stretched a net work of pitched cord, so as to protect the plants from external injury, and prevent the earth from being disturbed by mice or other vermin.

To each box should be fastened a canvass cover, made to go entirely over it, but so constructed as to be easily put on or off, as may be necessary, to protect the plants from the salt water, or winds, and sometimes from the sunshine. Strong handles should be fixed to the boxes that they may be conveniently moved.

During the voyage, the plants should be kept in a light, airy situation, without which, they will perish. They should not be exposed to severe winds, nor cold, nor for a long time to too hot a sunshine, nor to the spray of the salt water. To prevent injury from the saline particles with which the air is oftentimes charged at sea, (especially when the waves have white frothy curls upon them) and which on evaporation close up the pores of the plants and destroy them, it will be proper, when they have been exposed to them, to wash off the salt particles by sprinkling the leaves with fresh water.

The plants and seeds that are sown will occasionally require watering on the voyage, for which purpose rain water is best. If, in any special case, particular instructions on this point, or upon any other connected with the management of the plants during the voyage, be necessary, they should be made known to those having charge of the plants. But, after all, much will depend upon the judicious care of those to whom the plants may be confided during the voyage.

Plants of the succulent kind, and particularly of the Cactus family, should not be planted in earth, but in a mixture of dry sand, old lime rubbish and vegetable mould, in about equal parts, and should not be watered.

It may not be necessary, in every case, to observe all the precautions here recommended in regard to the putting up and transmission of seeds; but it is believed, that there will be the risk in departing from them, in proportion to the distance of the country from which the seeds are to be brought, and to the difference of its latitude, or of the latitudes through which they will pass on the voyage. It is not intended, however, by these instructions; to exclude the adoption of any other modes of putting up and transmitting seeds and plants, which are in use in any particular place, and which have been found successful, especially if more simple. And it is recommended, not only that the aid of competent persons be accepted in procuring and putting up the seeds and plants, but that they be invited to offer any suggestions in regard to the treatment of the plants during the voyage, and their cultivation and use afterwards.

Appendix 2

Alphabetized Version of William Elliot's List of Plants in the Botanic Garden of the Columbian Institute, Prepared in 1824 *

When reviewing the list, keep the following points in mind. The first column is a list of scientific plant names as they appear on Mr. Elliot's list, although spelling has been corrected or updated in several instances. This list is in alphabetical order, though the original was in random order. The second column is an update of the scientific nomenclature, where applicable. Common names are provided in the third column, when one is in generally accepted usage. Column four indicates the origin of each plant. The information used in columns two, three, and four was taken from *Hortus Third,* initially compiled by L.H. and E.Z. Bailey and revised by the Staff of the L.H. Bailey Hortorium (New York, NY,: MacMillan Pub. Co., 1977).

Scientific Plant Names as They Apeared on the Original List	Update of Nomenclature Where Applicable	Common Name	Origin
Abies picea	Abies alba	Silver Fir	Eur.
Acer campestre		Hedge Maple	Eur., Iran, Turkey, Caucasus
Acacia julibrissin	Albizia julibrissin	Silk Tree	Iran to Japan
Acer platanoides		Norway Maple	Eur., Asia
Acer tataricum		Tatarian Maple	Eur., Asia
Achillea ageratum		Sweet Yarrow	Eur.
Achillea setacea	(poss. Achillea millefolium)	Yarrow	Eur.
Agrimonia eupatoria		Agrimony	Eur., Asia, Afr.
Agrostemma coronaria	Lychnis coronaria	Mullein Pink	Afr., Eur.
Ailanthus glandulosa	Ailanthus altissima	Tree-of-Heaven	China
Alcea rosea (cv. name illegible)		Hollyhock	Asia Minor
Alchemilla vulgaris		Lady's-Mantle	Eur.

* Peter Force Papers, Manuscript Division, Library of Congress.

Scientific Plant Names as They Apeared on the Original List	Update of Nomenclature Where Applicable	Common Name	Origin
Allium cepa rubra		Onion	
Althea rosea sinensis	Alcea rosea	Hollyhock	Asia Minor
Anethum graveolens		Dill	Asia
Antirrhinum majus		Snap-dragon	Medit. Region
Artemisia absinthium		Sweet Yarrow, Absinthe, Wormwood	Eur.
Atriplex hortensis		Orach	Asia
Atriplex hortensis 'Rubra'		Orach	Asia
Atropa belladonna		Belladonna	Eurasia, Afr.
Balsamita (sp. name illegible)	Chrysanthemum balsamita	Costmary	Eur., W. Asia
Beta aurea	poss. Beta vulgaris cv.	Beet	
Betonica officinalis	Stachys officinalis	Betony	Eur., Asia
Brassica napus (cv. name illegible)		Rape, Cole, Mustard	
Brassica oleracea		Wild Cabbage	Eur.
Brassica oleracea (cv. name illegible)		Cabbage	Eur.
Broussonetia papyrifera		Paper Mulberry	Asia to Polynesia
Cacalia (sp. name illegible)	Emilia sp.		
Campanula medium		Canter-bury-Bells	Eur.
Chrysanthemum (sp. name illegible)			
Clematis flamula			Medit. Region to Iran

Scientific Plant Names as They Apeared on the Original List	Update of Nomenclature Where Applicable	Common Name	Origin
Clematis (sp. name illegible)			
Cneorum trioccon		Spurge Olive	Eur.
Colutea arborescens		Bladder Senna	Eur.
Coriandrum sativum		Coriander	Eur.
Cornus mascula	Cornus mas	Cornelian Cherry	Eur.
Cornus sanguinea		Blood-Twig Dogwood	Eur.
Crataegus amelanchier			
Crataegus aria	Sorbus aria	White Beam	Eur.
Crataegus torminalis	Sorbus torminalis	Wild Service Tree	Eur., Afr., Asia Minor
Crataegus sp.			
Cucurbita		Gourd	W. Hemis.
Curcurbita (sp. name illegible)			
Cynoglossum officinale		Hound's Tongue	Eur.
Cytissus laburnum	Laburnum anagyroides	Golden-Chain	Eur.
Eryngium maritimum		Sea Holly	Eur.
Echium vulgare		Blueweed	Eur., Asia
Euonymus europaea		European Spindle Tree	Eur., Asia
Euonymus latifolia			Algeria, Eur. to Iran

Scientific Plant Names as They Apeared on the Original List	Update of Nomenclature Where Applicable	Common Name	Origin
Ferula communis		Common Giant Fennel	Eur. to Syria
Fontanesia phillyreoides			Asia Minor
Genista juncea	Spartium junceum	Spanish Broom	
Hedysarum coronarium		French Honey-suckle	Eur.
Helianthus annuus		Common Sunflower	No. Amer.
Hibiscus syriacus		Rose-of-Sharon	Asia
Ilex aquifolium 'Balearica'		English Holly	
Ilex aquifolium 'Serratifolia'		English Holly	
Inula helenium		Elecampane	Asia
Ipomoea purpurea		Morning Glory	Eurasia, Afr.
Ligustrum (sp. name illegible)		Privet	
Ligustrum vulgare		Common Privet	Medit. Region
Lonicera alpigena		Honey-suckle	Eur.
Lunaria annua		Honesty, Money Plant	Eur., No. Amer.
Lycium barbarum		Matrimony Vine	Afr. to Iraq
Matricaria par-thenoides	Chrysanthemum parthenium	Feverfew	Eur. to Caucasus
Melissa officinalis		Lemon Balm	Eur.
Mespilus coccinea			

Scientific Plant Names as They Apeared on the Original List	Update of Nomenclature Where Applicable	Common Name	Origin
Mespilus lucida			
Mespilus oxyacantha			
Mespilus pyracantha	Pyracantha coccinea	Fire Thorn	
Morus alba		White Mulberry	Eur., No. Amer.
Morus italica			
Morus nigra		Black Mulberry	Asia
Nigella damascena		Love-in-a-Mist, Wild Fennel	Eur., Afr.
Ocimum basilicum		Sweet Basil	Old World Tropics
Oenanthe (sp. name illegible)			
Paliurus aculeatus	Paliurus spina-christi	Christ Thorn, Jerusalem Tree	Eur. to China
Pastinaca sativa (cv. name illegible)		Parsnip	Eurasia
Phaseolus lupinoides			
Phaseolus (sp. name illegible)		Bean	
Phaseolus (sp. name illegible)		Bean	
Phaseolus vulgaris		Kidney Bean	Trop. Amer.
Pinus sylvestris		Scots Pine	Eurasia
Pistacia terebinthus		Cyprus-Turpentine	Medit. Region
Pistacia trifolia			
Pisum sativum (cv. name illegible)		Garden Pea	

Scientific Plant Names as They Apeared on the Original List	Update of Nomenclature Where Applicable	Common Name	Origin
Prunus avium		Sweet Cherry	Eurasia
Prunus lusitanica		Portugal Laurel	Portugal, Spain to Canary Is.
Prunus mahaleb		Mahaleb St. Luci Cherry	Eur., Asia
Prunus padus		Bird Cherry	Eur. to Japan
Prunus padus (cv. name illegible)			
Quercus ilex		Holly Oak	Medit. Region
Rhamnus frangula		Alder, Buckthorn	Eur., Afr., Asia
Ribes rubrum		Red Currant	Eur., Asia
Ricinus communis		Castor Bean	Trop. Afr.
Robinia cragana			
Robinia pseudoacacia		Black Locust	U.S.A.
Robinia (sp. name illegible)			
Rosa cinnamomea		Cinnamon Rose	Eurasia
Rudbeckia amplexicaulis	Dracopis amplexicaulis	Coneflower	U.S.A.
Rumex patientia		Monk's Rhubarb	Eur.
Sambucus nigra		European Alder	Eur., Afr., Asia
Saponaria officinalis		Bouncing Bet, Soapwort	Eur., Asia
Sedum telephium		Orpine	Eur. to Japan

Scientific Plant Names as They Apeared on the Original List	Update of Nomenclature Where Applicable	Common Name	Origin
Solidago virgaurea		European Goldenrod	Eur., Asia, Afr.
Serratula alata			
Sorbus hybrida		Mountain Ash	Scandinavia
Staphylea pinnata		European Bladdernut	Eur.
Styrax officinalis		Snowbell, Storax	Balkans to Israel
Syringa vulgaris 'Alba' and 'Purpurea'		Common Lilac	
Teucrium botrys			
Ulmus campestris	Ulmus carpinifolia	Smooth-Leaf Elm	Eur., Afr., Asia
Verbena officinalis		Vervain	U.S.A.
Xeranthemum annuum		Immortelle	Eur.

APPENDIX 3

EXPERIMENTAL GARDEN OF THE NATIONAL INSTITUTE, 1844

A further elaboration of the Patent Office greenhouse is provided by an article written by a visitor to the facility in 1844.*

. . . The ship Vincennes of the Exploring Expedition, arrived home in the spring of 1842, bringing Messrs. Pickering and Breckenridge, [sic] the botanists attached to the Expedition, who brought with them, in addition to great quantities of seeds, bulbs & upwards of one hundred species of live plants. . . .

We were fortunate in finding our correspondent, Mr. Breckenridge, at home; and we had the pleasure of looking over the plants with him. Many of the species are quite rare, and now introduced for the first time. The following is a list of the more prominent plants:—

Aeschynanthus new sp., Ae. grandiflorus, Clerodendron sp., C. speciosissimus, Crinum amabile, Mimosa sensitiva (true,) a small shrub, Gardenia Thunbergia, Strelitzia spathulata, elongata and juncea, Ruellia sp., Agati grandiflora, Casuarina indica, or spear tree, from the South Sea Islands. Arduina grandiflora, Laurus sp. from California, Diosma sp., Genista sp., from the Canary Isles, Phlomis Leonurus, California rose, single, Oxalis elongatus, Babiana rubro caerulea, and many other bulbs, amaryllises & crinums.

In addition to these, of which there were in some instances, duplicate plants, Mr. Breckenridge had collected many fine plants together, and as they were in good health, the collection presented a very fine appearance. Another season, under his attentive care, we may look for a better development of the habits and character of many of the more rare and tropical species. Mr. Breckenridge will also, by that time, have multiplied many of the plants, to such a degree, that they may, if such is the intention of government, be distributed among nurserymen.

Of the seeds brought home, a larger part, we believe, lost their vegetative powers. Many of the seeds of Pines, of

---

*C.M. Hovey, "Experimental garden of the National Institute," Magazine of Horticulture, v. 10, March 1844, pp. 81-82.

California, such as the P. Lambertiana have been distributed, and in some instances, have grown; the ericas, from the Cape, have also proved good; but the greater portion of miscellaneous seeds, collected at the various places where the Expedition touched, have not vegetated even under the best care; at least such has been our experience, and the experience of many of our friends.

From the garden, we visited the institute, and examined some of the beautiful ferns which form part of the immense herbarium, collected by Messrs. Pickering, Breckenridge and Rich,—but no arrangement of the collection has yet been made, though Mr. Rich has long been at work, and the specimens were piled up in the sheets just as they were dried. We trust that measures will be taken by the Institute, to have all the specimens properly put up in good paper, ticketed and arranged according to the Jussieuian or Natural system and a complete and correct list of the entire collection published.

APPENDIX 4

MEMORIAL TREES PLANTED ON THE GROUNDS OF THE FIRST U.S. BOTANIC GARDEN.*

1. *The Crittenden Memorial Tree, a mossy-cup oak (Quercus macrocarpa)*

    Planted in 1863 by the Hon. J.J. Crittenden. Located near to and south of the east gateway entrance to the Garden. The acorn for this and a companion tree, planted at the same time by Mr. Robert Mallory (a pesonal friend of Senator Crittenden) were brought from Kentucky by these two gentlemen. Mr. Mallory's tree was planted, prior to its incorporation in the Botanic Garden, on the towpath of the old Washington, D.C., Canal and was located in the western section of the Garden.

2. *The Garfield Memorial Tree, referred to as an acacia, actually a silk tree or mimosa (Albizia julibrissin)*

    Planted along the walk near the south entrance to the large conservatory. A small seedling branchlet of this tree was placed on the coffin of President Garfield during the funeral ceremonies by a member of the Masonic fraternity. After the burial, the seedling was brought to Washington and planted on the Botanic Garden grounds. (Near this tree, on the opposite side of the walk, an acacia tree was planted as a memorial to the late General Albert Pike, for so many years the central figure of the Masonic fraternity in the United States.)

---

*Information on the Garden's memorial trees was compiled from two sources: Hearings before the Committee on Public Buildings and Grounds, House of Reps., Seventy-Fourth Congress, First Session on House Res. #221, May 28, 1935, Washington, D.C., U.S. Govt. Printing Office, 1935, pp. 25-26; and a letter to Frederick Law Olmsted, dated April 12, 1934, from the Architect of the Capitol, Botanic Garden, Files, Records of the Architect of the Capitol. Earlier information on trees to be found on the garden grounds was compiled by Frederick Law Olmsted in the Annual Report of the Architect of the United States Capitol, Washington, D.C., U.S. Govt. Printing Office, 1882, p. 10.

3. *The Beck Memorial Tree, an American elm (Ulmus americana)*

    Planted by Senator Beck of Kentucky south of the east gateway entrance to the Garden, near the Crittenden tree. This specimen was of special interest because it was propagated from the roots of the Washington elm, which once grew on the Capitol grounds.

4. *The Alexander Shepherd Memorial Tree, an American elm (Ulmus americana)*

    Planted in close proximity to the Beck Memorial Tree. It too was propagated from the roots of the Washington elm, formerly on the Capitol grounds.

5. *The J.W. Forney and Edwin Forrest Memorial Trees, two bald cypresses (Taxodium distichum)*

    Planted on opposite sides of the walk at the southern entrance to the Garden.

6. *The Conger Memorial Tree, a Spanish red oak (Quercus falcata)*

    Planted by the late Senator Conger during his term of service in Congress. This tree was located south of the east gateway entrance to the Garden.

7. *The Hayes Memorial Tree, an oak (Quercus leana)*

    Planted near the west gateway entrance to the Garden by Mr. Hayes during his term in Congress as Chairman of the Library Committee. This rare native oak was reportedly one of very few specimens found growing in the United States at that time.

8. *The Palmer Memorial Tree, a Japanese walnut (probably Juglans ailantifolia)*

    Planted by Mr. Palmer, the public printer, near the west gateway entrance to the Garden, north of the Hayes Memorial Tree.

9. *The Bingham Memorial Tree, a European hornbean (Carpinus betulus)*

    Planted bordering the south walk of the Garden, between First and Second Streets. (The Charles Summer Tree, formerly one of the most ornamental trees on the Capitol grounds and for the preservation of which Senator Summer made an eloquent plea in the U.S. Senate, was also a tree of this species.)

10. *The Stewart Memorial Tree, a red oak (Quercus rubra)*

    Planted by Senator Stewart of Michigan near the east gateway entrance to the Garden.

11. *The Torrey Memorial Tree, a swamp oak (Quercus bicolor)*

    Planted by Doctor Torrey south of the east gate.

12. *The Hoar and Everts Memorial Trees, referred to as cedars of Lebanon but actually atlas cedars (Cedrus atlantica 'Glauca')*

    Planted in close proximity to one another by Senators Hoar and Everts along the south walk of the Garden.

13. *The Holman Memorial Tree, a Crimean fir (Abies, species unknown)*

    Planted by the distinguished Member of Congress from Indiana. It was located on the lawn bordering the walk leading to the west door of the large conservatory.

14. *The Dan Voorhees Memorial Tree, a sycamore (Platanus orientalis)*

    Planted near the west gateway to the Garden by Mr. Voorhees during his term in Congress.

15. *The Morrill Memorial Trees, two winged elms (Ulmus alata)*

    Planted by Lot M. and Justin S. Morrill during their terms in the Senate. These trees were located near the western end of the central walk of the Garden.

Supposedly, they were the only specimens of the winged elm growing in the city at that time.

16. *The Blackburn Memorial Trees, a yulan magnolia (Magnolia conspicua, syn: Magnolia heptapeta) and a Japanese elm (Ulmus davidiana var. japonica)*

    Supposedly planted near the large conservatory in the center of the Garden.

17. *The O.R. Singleton Memorial Tree, a European cut leaved linden (Tilia x vulgaris lacinata, syn: T. x europaea var.)*

    Planted on the main walk near the center of the Garden. This was reportedly the only specimen of cut leaved linden growing in Washington at that time.

18. *The James H. Pierce Memorial Tree, a swamp white oak (Quercus bicolor)*

    No additional information provided.

19. *The Secretary Bayard Memorial Tree, an English oak (Quercus robur)*

    Planted near the east gateway entrance to the Garden.

20. *A Chinese oak (Quercus serrata, syn: Quercus acutissima)*

    A friend of the Hon. Charles A. Dana was travelling in China and obtained a numer of acorns under a tree growing by the grave of Confucius. He brought them to America and gave them to Dana, who planted a number of the acorns on the grounds of his home. One of the acorns was presented to Mr. William R. Smith by Mr. Falconer, a horticulturist, who was at that time superintendent of Mr. Dana's extensive grounds at Glen Cove, Long Island. Smith had the acorn planted south of the greenhouses near Maryland Avenue and Second Streets, Southwest.

21. *The Hanna and Dick Memorial Trees, two Japanese umbrella pines (Sciadopitys verticillata)*

    Planted by Senator Charles Dick on the lawn bordering the south walk near the entrance gateway to the Garden.

Other trees of special interest including the Proctor Knot Tree, an English Oak (*Quercus robur*) planted to commemorate the settlement of Alabama claims; and two American elms (*Ulmus americana*) taken as seedlings from the historical Washington elm growing at Cambridge, Massachusetts.

Appendix 5

Species Included in "A Catalog of Plants in the National Conservatories," Prepared by William R. Smith in 1854*

Scientific Plant Name	Common Name	Origin
Acacias in variety	Acacia	
Agave americana	Century Plant	Mexico
Aleurites moluccana	Candle-Nut Tree	Asia
Aloe species		Old World (chiefly Afr.)
Annona cherimola	Cherimoya	Andes of Peru and Ecuador
Annona muricata	Soursop	Trop. Amer.
Antiaris toxicaria	Upas Tree	Trop. Afr. and Asia to Philippine Is. and Fiji Is.
Araucaria heterophylla	Norfolk Island Pine	Norfolk Island
Ardisia species	Ardisia	
Aristolochia grandiflora	Pelican Flower	Caribbean
Azaleas in variety	Azalea	
Babiana species	Baboon Flower	So. Afr.
Banksia robur		Queensland, New So. Wales
Banksia serrata		Queensland, New So. Wales
Banksia species		
Brunsvigia species		
Cactus in variety		
Calathea zebrina	Zebra Plant	Brazil
Callistemon citrinus	Crimson Bottlebrush	Australia
Callistemon linearifolius	Bottlebrush	New So. Wales

*William R. Smith, "A Catalogue of Plants in the National Conservatories," A Popular Catalogue of the Extraordinary Curiosities in the National Institute Arranged in the Building Belonging to the Patent Office, Washington: Alfred Hunter, 1854, p. 64.

Scientific Plant Name	Common Name	Origin
Callistemon species	Bottlebrush	
Camellia japonica in variety	Camellia	
Camellia sinensis	Tea Plant	Asia
Ceratonia siliqua	St. John's-Bread	Medit. Region
Cinnamomum camphora	Camphor Tree	China, Japan, Taiwan
Citrus limon	Lemon	Asia
Coffea arabica	Coffee	Trop. Afr.
Cycas revoluta	Sago Palm	So. Japan, Ryukyu Is.
Daphne odora	Winter Daphne	China
Dracaena sp.	Ti	Old World Tropics
Encephalartos horridus	Cycad	So. Afr.
Ericas in variety	Heath	
Euphorbia pulcherrima	Poinsettia	Cent. Amer., Trop. Mex.
Ficus elastica	India Rubber Tree	Nepal to Assam and Burma
Furcraea foetida	Mauritius Hemp	So. Amer.
Hedychium garderanum	Kahili Ginger	India
Hypoxis stellata	Star Grass	So. Afr.
Illicium floridanum	Purple Anise	Fla. to La.
Ixia sp.	Corn Lily	So. Afr.
Jubaea chilensis	Chilean Wine Palm	Chile
Kalanchoe pinnata	Air Plant	
Latania lontaroides	Red Latan	Reunion Is.
Laurus nobilis	Laurel	Medit. Region
Lawsonia inermis	Henna	Afr., Asia
Leptospermum scoparium	Tea Tree	New Zeal., Tasmania

Scientific Plant Name	Common Name	Origin
Mangifera indica	Mango	Asia
Michelia figo	Banana Shrub	China
Mimosa pudica	Sensitive Plant	Trop. Amer.
Musa acuminata	Edible Banana	Asia
Musa coccinea	Banana	Indochina
Musa sp.	Banana	Asia
Nepenthes destillatoria	Pitcher Plant	
Nerine sp.		
Orchids in variety including:		
Peristeria elata	Holy Ghost Flower	
Epidendrum sp.	Buttonhole Orchid	
Oncidium papilio	Butterfly Orchid	
Vanilla sp.	Vanilla	
Stanhopea sp.		
Cattleya sp.	Corsage Orchid	
Dendrobium sp.		
Pandanus sp.	Screw Pine	Old World Tropics
Passiflora laurifolia	Yellow Grandilla, Water Lemon	W. Indies to Brazil and Peru
Persea americana	Alligator Pear, Avacado	Trop. Amer.
Phoenix dactylifera	Date Palm	Asia and Afr.
Phormium tenax	New Zealand Flax	New Zealand
Phyllanthus sp.		
Piper nigrum	Black Pepper	India and Ceylon
Platycerium grande	Staghorn Fern	Australia
Pyrostegia venusta	Flame Vine	Brazil, Paraguay
Saccharum officinarum	Sugarcane	Asia
Strelitzia reginae	Bird-of-Paradise	Afr.
Yucca aloifolia	Spanish-Bayonet	No. Amer.

APPENDIX 6

KEIM'S DESCRIPTION OF THE BOTANICAL COLLECTION, 1875*

BOTANICAL COLLECTION.—The first collection of plants in this National Conservatory was brought to the United States by the Exploring Expedition to the Southern Hemisphere, 1838–42, commanded by Captain (Rear Admiral) Charles Wilkes. The collection was first deposited in the Patent Office, but in 1850 was removed to the Botanical Garden. Some of the plants are still living, and a large share of the present collection are the descendants of those brought back by the Wilkes Expedition. A few have furnished representatives for many of the principal conservatories of the United States and Europe.

The disposition of the collection is according to a geographical distribution. The strictly tropical plants occupy the centre Conservatory, and those of a semi-tropical nature, requiring protection and lying towards the N. pole, are placed in the W. range and wing; and all indigenous to countries lying towards the S. pole are in the E. range and wing.

The *Centre Building* or *Rotunda*, temperature 80°, contains a fine variety of the majestic palms, called by Martius the princes of vegetation, and of which there are 300 kinds, the most prominent being here represented. The most interesting in the collection is the palm tree of Scripture, familiarly known as the date palm. Jericho, the City of Palms, was so called from the numbers of this tree growing in its vicinity. It was recommended to be used by the Jews in the Feast of Tabernacles. In Arabia, Egypt, and Persia it supplies almost every want of the inhabitants. The fruit is used for food, the leaves for shelter, the wood for fuel, and the sap for spirituous liquor. It matures in 10 years and then fruits for centuries, bearing from 1 to 300 cwt. at a time. Among the Arabs the pollen dust is preserved from year to year, and at the season of impregation of the pistils or female flowers a feast called "Marriage of the Palms" is held. It is a singular historical fact, that the date palm of Egypt bore no fruit in the year 1800, owing to the presence of the French army in the country, which prevented the annual marriage feast.

Among the other plants in this portion of the Conservatory are the fan, royal, ratan, sago of Japan and China, Panama hat,

---

*DeB. Randolph Keim, Keim's Illustrated Hand-Book of Washington and Its Environs: A Descriptive and Historical Hand-book to the Capital of the United States of America, Washington: DeB. Randolph Keim, 1875, pp. 41-42.

oil, wine, coco de Chili, sugar, and cradle palms; the East India bamboo; the tree fern from New Zealand; astrapea, from Madagascar; screw pine of Australia, with its cork-screw leaves and roots in mid air; the cinnamon of Ceylon; maiden's hair fern; mango, a delicious fruit of the West Indies; and banana, that most prolific of all plants; the great stag and elkhorn ferns from Australia, (very fine specimens,) and the dumb cane of South America. The sap of the root of the latter will take away the power of speech. Humboldt, during his explorations in South America, was eight days speechless from tasting it. The outer circle of the rotunda is devoted to the smaller tropical plants.

The *E. range,* temperature 50°, and *wing,* 40°, are devoted more particularly to the plants of the South Sea Islands, Brazil, Cape of Good Hope, Australia, and New Holland. The principal specimens are the tree fern of New Zealand; the aloe and the Caffre bread tree from the Cape of Good Hope; the India rubber, the passion flower, the caladium, of Brazil; Norfolk Island pine of Australia, one the most beautiful and largest-growing trees in the world; the queen plant, or bird of paradise flower, from its resemblance to the plume of that bird; the tutui, or candle-nut tree, from the Society Islands, the nut being used by the natives for lighting their huts; the coffee plant, and several varieties of cactus.

To the *W. range* and *wing,* temperature same as E., the plants of China and Japan, the East and West Indies, and Mexico are assigned. The most notable plants here are the cycadaceae, of the East Indies, the largest in the country; the four-century plant; the camellia japonica, or Japan rose; the lovely lily of Cuba; the historic *papyrus antiquorum,* or paper plant, of Egypt; the tallow and leechee trees of China; the guava, a delightful fruit of the West Indies; the vanilla of Mexico, the species which furnishes the aromatic bean; the black pepper from the East Indies; the sugar cane, the cheramoyer, or custard apple, and cassava of the West Indies; the sensitive and the humble plants; the American aloe, or century plant, of Mexico; the camphor tree from Japan; the tea plant; the papay, an Oriental tree, which has the property of rendering the toughest meat tender; a plant of the *adansonia digitata,* or monkey bread, which grows on the banks of the Senegal, and reaches the enormous circumference of 100 ft. They are supposed to attain the age of 5,000 years. They have many uses. Humboldt pronounces them the oldest organic monuments of our planet. There is also a specimen of the carob tree of Palestine, sometimes called St. John's bread. The pulp around the seed is supposed to have been the wild honey upon which St. John fed in the wilderness. There are other interesting specimens of the vegetable kingdom, including a pleasing variety of climbing plants. The arrangement of the exotics in the Central Conservatory presents the appearance of a miniature tropical forest, with its luxuriant growth of tree and vine. Until recently the Conservatory was in possession of a specimen of the bohan upas tree, of which such fabulous stories have been told. Each wing of the Conservatory is supplied with a fountain. In the W. range is a vase, brought from St. Augustine, Florida, and taken from the first house built on the North American continent within the present limits of the United States. A fine specimen of maiden's hair fern grows in the vase.

The outside conservatories are generally used for propagation. One, however, is specially devoted to camellia japonica, and another to that curious growth, the orchids or air plants. The botanical collection received some valuable contributions from the expedition of Commodore Perry to Japan. The supply is kept up by propagation and purchase, and at rare intervals by scientific or exploring expeditions of the United States.

APPENDIX 7

PLANTS COLLECTED DURING THE PERRY EXPEDITION OF 1852 to 1855 AND DONATED TO THE BOTANIC GARDEN.*

1. Yellow or Tea Roses
2. China Monthly Dark Roses
3. China Monthly Light Red Roses
4. China Monthly Pink Roses
5. Can "Fas" or Flower
6. Blue Magnolia
7. Qui-"Fas"
8. China Grafted Lilac Roses
9. Koco or Small Magnolia Flower
10. China Yellow Aromatic
11. Lary Hymerucalus
12. Logan
13. Guavas
14. Loquat
15. Cutlan Apple
16. Sweet Whampee
17. Pumbalos
18. Mangoes
19. Large Mandarin Orange "Even" Skin
20. Small Mandarin Orange "Even" Skin
21. China Mandarin Orange Hard Skin
22. Cumquats
23. Large Yellow Persimmon
24. Large Round Sour Apple
25. Large Round Red Persimmon
26. Small Round Red Persimmon
27. Double Myrtle
28. Dates
29. Large Long Rose Apple
30. Small Long Rose Apple
31. Lychee
32. Papaya
33. China Common Oranges Hard Skin
34. Very Fine Brown Skin Mandarin Oranges
35. Black Tea
36. Kar-Diempet (Blue Lily)
37. Red Double Head Star Lily
38. China Red Lily
39. China Yellow Lily
40. Single Head White Lily
41. Cymbedium
42. Nondescript White Small Plants
43. Lemon Grass
44. Peunian Cryuum
45. Small Hymanicalus
46. Benjamin Flower
47. Blue Magnolia
48. Round Rose Apple
49. China Dates

---

*A.B. Cole, A Scientist with Perry in Japan: The Journal of Dr. James Morrow, Chapel Hill: University of No. Carolina Press, 1947, pp. 263-65.

Appendix 8

Plants Growing Inside the Conservatory and on the Grounds of the First Botanic Garden During the George W. Hess Years (1913–1934)*

CONSERVATORY PLANTS

Scientific Plant Name	Common Name	Origin
Acokanthera oblongifolia	Wintersweet	Sw. Afr.
Agave sp.	Century Plant	No. & So. Amer.
Andropogon sp.	Bluestem	
Anthurium sp.	Tailflower	Trop. Amer.
Araucaria araucana	Monkey-Puzzle Tree	Chile
Araucaria columnaris	New Caledonian Pine	New Caledonia, New Hebrides
Araucaria heterophylla	Norfolk Island Pine	Norfolk Is.
Arenga pinnata	Sugar Palm	Malay Arch.
Artocarpus altilis	Breadfruit	Malay Pen.
Asimina triloba	Pawpaw	No. Amer.
Asplenium nidus	Bird's-Nest-Fern	Trop. Asia, Polynesia
Attalea sp.		Trop. Amer.
Calamas Sp.	Rattan Palm	
Camellia sinensis	Tea Plant	Asia
Carica papaya	Papaya	Trop. Amer.
Carludovica palmata	Panama-Hat Palm	Cent. Amer. to Bolivia
Caryota urens	Fishtail Palm	India, Ceylon, Malay Pen.
Cinchona sp.	Quinine	So. Amer.
Cinnamomum camphora	Camphor Tree	China, Japan, Taiwan

*List compiled by Karen D. Solit from articles on the garden which appeared in the Washington Star between the years 1914 and 1917.

Scientific Plant Name	Common Name	Origin
Cinnamomum zeylanicum	Cinnamon	Ceylon, India
Clivia sp.	Kaffir Lily	So. Afr.
Codiaeum hybrids	Crotons	
Coffea arabica	Coffee	Trop. Afr.
Cycas revoluta	Sago Palm	So. Japan, Ryuku Is.
Desmodium motorium	Telegraph Plant	Trop. Asia
Dicksonia antarctica	Tasmanian Tree Fern	Aust., Tasmania
Dieffenbachia maculata	Dumb Cane	Cent. & So. Amer.
Dionaea muscipula	Venus's-Flytrap	No. & So. Carolina
Eriobotrya japonica	Loquat	Asia
Euphorbia longan	Longan	India
Euphorbia milii	Crown-of-Thorns	Madagascar
Ferns in variety		
Ficus lyrata	Fiddle-Leaf Fig	Trop. Afr.
Ficus nekbudu	Kaffir Fig	Trop. Afr.
Ficus pumila	Creeping Fig	Asia
Fittonia vershaffeltii	Mosaic Plant	Columbia to Peru
Ilex paraguariensis	Paraguay Tea	Paraguay, Brazil, Argentina
Laurus nobilis	Laurel	Medit. Region
Lodoicea maldivica	Double Coconut	Seychelles Is.
Manihot esculenta	Tapioca	Brazil
Mimosa pudica	Sensitive Plant	Trop. Amer.
Monstera deliciosa	Swiss-Cheese Plant	Mex., Cent. Amer.
Musa sp.	Banana	Asia
Nephrolepis exaltata cvs.	Boston Ferns	
Orchids in variety		
Pandanus sp.	Screw Pine	Old World Tropics
Pelargonium sp.	Scented Geraniums	
Phoenix sp.	Date Palm	Afr., Asia

Scientific Plant Name	Common Name	Origin
Pimenta dioica	Allspice	W. Indies, Cent. Amer.
Piper betle	Betel	Malay Pen. to India
Pistia stratiotes	Water Lettuce	Tropics
Platycerium alcicorne (p. vassei)	Staghorn Fern	Australia
Ravenala madagascariensis	Traveler's Tree	Madagascar
Rhopalostylis sapida	Nikau Palm	New Zealand
Sabal palmetto	Cabbage Palm	No. Carolina to Fla., Bahama Is.
Sarracenia flava	Yellow Pitcher Plant	No. Amer.
Strelitzia reginae	Bird-of-Paradise	So. Afr.
Theobroma cacao	Cacao	Cent. & So. Amer.
Thrinax sp.	Thatch Palm	
Zingiber sp.	Ginger	Trop. Asia

OUTDOOR PLANTINGS

	Annuals in variety	
	Bulbs in variety	
Carpinus sp. (supposedly planted by Abraham Lincoln)	Hornbeam	No. Hemisphere
Forsythia sp.	Forsythia	
Genista sp.	Broom	
Helleborus orientalis	Lenten Rose	Eur., Asia
Liquidambar styraciflua	Sweet Gum	No. & So. Amer.
Magnolia kobus		Japan
Magnolia quinquepeta		China
Magnolia x soulangiana	Saucer Magnolia	
Magnolia x soulangiana 'Norbertii'		

Scientific Plant Name	Common Name	Origin
Magnolia stellata	Star Magnolia	Japan
Viola	Pansies in variety	
Paeonia	Peonies in variety	
Poncirus trifoliata	Trifoliate Orange	China
Quercus ilex	Holly Oak	Medit. Region
	Rock garden plants in variety	
Prunus tomentosa	Nanking Cherry	Asia
Spiraea japonica	Japanese spirea	Asia

APPENDIX 9

Plants for Congressional Distribution Through the Botanic Garden, 1930

UNITED STATES BOTANIC GARDEN
OFFICE OF THE DIRECTOR
Washington, D.C.

PLANTS FOR CONGRESSIONAL DISTRIBUTION, 1930

Botanical name	Common name	Habitat
TREES		
Acer desycarpum	Silver maple	United States.
Citrus trifoliata	Hardy orange	China.
Gleditschia triacanthus	Honey locust	United States.
Ginkgo biloba	Maiden hair tree	China.
Gymnocladus Canandensis	Kentucky coffee	United States.
Laurus Camphora	Camphor tree	Ceylon. For South or conservatory.
Salix Babylonica	Weeping willow	Asia.
Salix Caprea	Pussy willow	Europe.
Quercus rubara	Red oak	Do.
Ulmus Americana	American elm	United States.
SHRUBS		
Abelia grandiflora	Rock abelia	China.
Altheas in variety	Rose of Sharon	Europe and Asia.
Artimesia abrotanum	Southern wood; old man	Europe.
Aucuba Japonica	Gold dust plant	Japan.
Cornus Sanguinea	Red-stemmed dogwood	North America.
Cornus stolonifera	do	Do.
Cornus lutea	Yellow-stemmed dogwood	Do.
Caryopteris mastacanthus	Blue spirea	China.
Cydonia Japonica	Japanese quince	Japan.
Deutzia gracilis	White deutzia	China.
Deutzia gracilis rosea	Pink deutzia	Do.
Deutzia crenata	Double deutzia	Do.
Deutzia scabra, Pride of Rochester.	Double white deutzia	Garden hybrid.
Deutzia scabra	White deutzia	China.
Deutzia Lemoineii	Lemoine's deutzia	Do.
Euonymus radicans	Creeping euonymus	Japan.
Euonymus radicans variegata	Variegated creeping euonymus	Do.
Eleagnus Simonsii	Oleaster	Do.
Eleagnus longipes	do	Do.
Forestiera acuminata	Adelia	Do.

PLANTS FOR CONGRESSIONAL DISTRIBUTION, 1930

Botanical name	Common name	Habitat
SHRUBS—Continued		
Forsythia viridissima	Upright golden bell	China.
Forsythia intermedia	do	Do.
Forsythia suspensa	Weeping golden bell	Do.
Forsythia Fortuneii	Fortnee's golden bell	Garden hybrid.
Hibiscus Sinensis	Hibiscus	China. For South or conservatory.
Hibiscus Cooperii	Cooper's hibiscus	Do.
Hydrangea Hortensis	Hydrangea	Japan.
Jasminum multiflorum	Star jasmine	China. For South or conservatory.
Jasminum fruiticans	Yellow jasmine	Europe.
Lagerstroemia Indica rosea	Pink crêpe myrtle	China. Not hardy north of District of Columbia.
Lagerstroemia Indica alba	White crêpe myrtle	Do.
Lagerstroemia Indica lavendula.	Lavender crêpe myrtle	Do.
Ligustrum aureum	Golden privet	Garden hybrid.
Ligustrum Ibota	Privet	Japan.
Ligustrum buxifolia	do	Asia.
Ligustrum medium	do	Japan.
Ligustrum Regelainum	Regel's privet	Do.
Lonicera tatarica alba	White honeysuckle	Europe and Asia.
Lonicera tatarica rosea	Pink honeysuckle	Do.
Lonicera tatarica rubra	do	Do.
Lonicera fragrantissima	Fragrat bush honeysuckle.	China.
Malvaviscus arboreus	Achania	South America. For South or conservatory.
Podocarpus Japonica	Podocarpus	Japan.
Philleria Angustifolia	Jasmine box	Europe.
Philadelphus coronarius	Syringa; mock orange	Do.
Philadelphus grandiflorus	do	Do.
Philadelphus parviflora	do	Do.
Philadelphus hirsutus	do	Do.
Philadelphus Ignus	do	Do.
Philadelpus Souvenir De Billard.	do	Do.
Spirea Anthony Waterer	Red spirea	Garden hybrid.
Spirea Van Houteii	Bridal wreath	China.
Spirea Reevesiana	Reeve's spirea	Do.
Spirea Prunifolia	White spirea	Do.
Spirea Billardi alba	Billard's white spirea	Do.
Spirea Billardi rosea	Billard's pink spirea	Do.
Spirea Opulifolia	White spirea	Do.
Staphylea Bumalda	Bladder nut	Europe.

PLANTS FOR CONGRESSIONAL DISTRIBUTION, 1930

Botanical name	Common name	Habitat
SHRUBS—Continued		
Staphylca triflora	Bladder nut	Europe
Syringa vulgaris	Lilac	China and Japan.
Syringa Persica	Persian lilac	Asia.
Tamarix Gallica	Tamarix	Do.
Tamarix Amurensis	do	Do.
Vitex Agnus Castus	Chaste tree	Europe and Asia.
Viburnum in variety	Snowball	China and Japan.
Weigelia rosea	Pink weigelia	China.
Weigelia Amabilis	do	Do.
Weigelia Eva Rathka	do	Do.
Weigelia Candida	White weigelia	Do.
PLANTS FOR BORDERS AND ROCKERIES		
Calliopsis lanceolata	Tickseed	United States.
Coreopsis lanceolata	do	Do.
Chrysanthemum leucanthemum	Shaster daisy	Do.
Aquilegia hybrida	Columbine	Do.
Gaillardia grandiflora	Blanket flower	Garden hybrid.
Althea rosea	Hollyhocks	China.
Statice latifolia	Sea pink	Europe.
Stokesia cyanea	Cornflower aster	Garden hybrid.
Bellis perrenis	English double daisy	Europe.
Dianthus in variety	Hardy pinks	China.
Matricaria Capensis	Feverfew	Europe.
Physostegia Virginica	False dragon head	United States.
Cheiranthus linifolium	Wall flower	Europe.
Phlox paniculata	Summer phlox	United States.
Pentstemon Torreyii	Pentstemon	Do.
VINES FOR CREEPING AND CLIMBING FOR WALLS OR TRELLIS		
Ampelopsis hetrophylla		Asia.
Akebia quinata	Rapid growing vine	Japan.
Bignonia capreolata	Cross vine	United States.
Bignonia radicans grandiflora.	Trumpet creeper	Do.
Hedera helix	English or evergreen ivy	Europe.
Jasminum primulinum	Yellow climbing jasmine	Europe. For South or conservatory.
Jasminum grandiflorum	Italian royal or Spanish jasmine.	Do.
Lonicera Japonica	Honeysuckle	Japan.
Peuraria Thunbergiana	Kudzu vine	Do.

111

PLANTS FOR CONGRESSIONAL DISTRIBUTION, 1930

Plants for Bedding, Borders, and Window Boxes

Coleus in variety.
Creeping Coleus.
Achryanthus in variety.
Ageratum (blue).
Cuphea platycentra.
Peristrophe Angustifolia variegata.
Plumbago capensis.
Plumbago coccinea.

Anthericum variegatum.
Fuchsia.
Balm Scented Geranium.
Daisjes.
Lantana hybrida.
Creeping Lantana.
Mesembryanthemum.
Summer flowering annuals.

Tender Vines for Window Boxes, Urns, Etc.

Cissus Amazonica.
Passiflora trifuscata.
Senecio scandens (parlor ivy).
Ipomea tuberculata.

Ipomea Learii
Vinea Major
Ipomea grandiflora (moon vine).

NOTE.—Each Member's quota consists of 70 plants, including trees, shrubs, and bedding plants in variety. Each request for plants to be shipped out of the city, should be accompanied by an addressed document frank. This list is for the use of Members of Congress only.

GEO. W. HESS, *Director.*